A SON OF 'THE RED CENTRE'

Memoirs and Anecdotes of the
Life of Road Train Pioneer and Bush Inventor
of the Northern Territory of Australia

BY
KURT G. JOHANNSEN
(Edited by Daphne Palmer)

Published by Kurt Johannsen
3 Stephens Place, Morphettville, South Australia 5043
Phone/Fax (08) 8294 4981 Adelaide, or
Alice Springs, N.T. Phone (08) 8952 1233 Fax (08) 8952 2031

First published 1992
Revised edition 1993, Reprinted 1993, 1994, 1996, 1997, 1999

Note:
Both metric and imperial measurements have been used throughout this book due to the fact that much of the material deals with pre-1966 days before metric was officially introduced in Australia.

© Kurt G. Johannsen, 1992
All rights reserved. No part of this publication may be reproduced, stored in a retrieval system, or transmitted in any form or by any means, electronic, mechanical, photo-copying, recording or otherwise, without the express written permission of the author.

National Library of Australia Cataloguing-in-Publication entry.
Johannsen, Kurt G. (Kurt Gerhardt), 1915-
A son of the Red Centre

   Rev. ed
   Bibliography.
   Includes index.
   IBSN 0 646 13304 7 pbk.
   IBSN 0 646 13303 9 hbk.
   1. Johannsen, Kurt G. (Kurt Gerhardt), 1915-. 2. Road trains-Northern Territory-Anecdotes.
   3. Inventors-Northern Territory-Biography. 4. Stockmen-Northern Territory-Biography.
   5. Frontier and pioneer life-Northern Territory-Anecdotes. I.Palmer, Daphne. II. Title.
   994.29092

Artwork & Layout prepared by Graphic Mac
Printed by Hyde Park Press,
4 Deacon Avenue, Richmond, South Australia 5033

Book design by Daphne Palmer

PHOTOGRAPHS:
Front Cover: *Kurt Johannsen and 'Bitzer' Mulga Express Mark I outside the Centralian Store,*
                     Alice Springs, N.T. 1936
Back Cover: *Kurt Johannsen and Mulga Express Mark IV and the wood-gas producer, 1987*
                     (Photo by Carmel Sears, courtesy 'Centralian Advocate').

Oh how I love the bush!
Camping in the shade of weeping ironwood trees
The sweet smell of flowering wattle
The whispering breeze through the desert oaks.
I love walking out in a summer shower
Inhaling earthen smells so pure
And watching buds burst out in flower.
I love the wide open spaces
And mountain ranges with such fascinating places
The whistling call of the kite hawks
Circling and gliding above
The red desert sandhills glowing in the evening light;
I love the flashing lightning at night
Painting pictures in the clouds so bright
The distant rumble of thunder
Then the crashing silence of the star-lit night
And waking to a symphony of birds' delight
When another day is born.

Kurt G. Johannsen, 1992

*Dedicated to the
memory of my
pioneering parents of
'The Red Centre'*

ACKNOWLEDGEMENTS

For transcribing the original stories from audio tapes, my thanks to Daphne Palmer, Trish Lonsdale and to my daughter, Paula Johannsen Hall. Thanks to Daphne Little, M.B.E., for help in proof-reading and to Gil Porter for advice on typesetting. My special thanks to Daphne Palmer for editing, book design and organizing my story.

For help with details and dates my appreciation and thanks to many family members and friends, including Russell Goldflam from ABC in Alice Springs and Judy Robinson. For photographs I express my gratitude and thanks to Judy Robinson (page 42), Helene Burns (pages 85, 87 and 88), Bryan Fitzpatrick (page 227), Graham Mudge who worked for me at Molyhil in 1972-3 (pages 229 and 232), Kath Greatorex, and to my sisters Trudy Hayes and Mona Byrnes, O.A.M. Many of the photographs of early days were taken by Trudy or my father. Thanks also to Carmel Sears from the *Centralian Advocate* for permission to use the photographs on pages 188, 240, 242 and Back Cover, taken in 1987, and to the State Library of N.T. (N.T. Collection) for permission to use a photograph on page 47 taken by the RESO party in 1927.

Thanks also to Jose Petrick for permission to use information contained in unpublished writings of the *Background and Family History of the Johannsen Family* compiled by her and other family members and also information contained in her *History of Alice Springs Through Street Names*, 1989. My appreciation also for permission to use headlines, quotations and information from the following newspapers: *Centralian Advocate*, Alice Springs and *The Advertiser*, Adelaide, South Australia. Thanks to John Maddock for permission to quote from his book *A History of Road Trains of the Northern Territory, Kangaroo Press, 1988*. And finally, thanks to my friends and family who encouraged me to start this project in the first place for the information and pleasure of future generations.

KURT G. JOHANNSEN
1992

NOTE FROM THE AUTHOR

If I have made errors or omissions in the names, dates or events recounted in this autobiography, my sincere apologies. There has been no deliberate intention on my part to offend any living person in the telling of my tales.

FOREWORD

In the late 1800's and early 1900's quite 'ordinary', humble, hard-working men and women set off into the unknown in the Northern Territory and lived remarkable, extraordinary, courageous lives. We are often astounded at what they endured and did when we look back over the past few decades of their achievements and heartbreaks in that beautiful, wild and often harsh and uncompromising land.

Among those families who were pioneers of 'The Red Centre' of the Territory in the early 1900's were Gerhardt and Ottilie Johannsen and their six children. Their descendants are now many and can be found around Australia, with a good number of them still living in the Territory and contributing to its development. Two of Kurt's sisters, Mona Byrnes and Myrtle Noske are well-known artists living in Alice Springs; Mona is also active in the Historical Preservation Society.

Kurt Gerhardt Johannsen, born and bred in 'The Red Centre', is a quiet, modest man with a knack for improvising, fixing things and inventing solutions to the many problems he has encountered in his working life during the infancy of road transport, as well as in his mining and other adventures. He is numbered among BP's 'Quiet Achievers' 1988, and his name can be found in one of the plaques placed along the Esplanade in Darwin, Northern Territory.

These memoirs, first recounted onto audio tapes by Kurt in 1987, following heart surgery in December, 1986, tell many of his adventures, mishaps and achievements and reveal some of his fortitude, integrity, humour and intelligence. He is a visionary with ideas that were often ahead of his time. He is also what modern psychologists call a 'lateral thinker', who, without benefit of much formal education, has acquitted himself well in the classroom of Life as a mechanic, engineer, inventor, miner, builder, bushman and 'Jack-of-all-trades'.

It has been my privilege to assist Kurt in compiling these memories of his action-filled life.

Daphne Palmer, B.A., B.Ed., 1992

CONTENTS

PART I–DEEP WELL (1915-1928)

Chapter

1. The Hardy Souls of Pioneers 1
2. Living Near the Desert Oaks 8
3. Visitors and Bush Folk 19
4. Trips South 25
5. Interlude at Hermannsburg (1922–1924) 32
6. Hard Times Back at Deep Well 36

PART II—PRE-WAR DAYS (1928–1939)

Chapter

7. Our Move to the Alice 49
8. First Work and Early Romance 56
9. First Mails and Mining 63
10. Trips for Tourists and Terrible Tracks 73
11. Sam Irvine and Other Characters 77
12. Anthropologists to Mount Liebig 84
13. Stud Rams to Kerr's 89
14. Foy's Expedition to the Western Desert 91
15. A Holiday Down South in the 'Big Smoke' 95
16. Aulgana Mica Mines 100
17. Rough Roads 103
18. Building 'Bitzers' 107
19. My Wettest Mail Runs 113
20. More Colourful Characters 119
21. A Disastrous Trip 122
22. Prospecting and Mining 125

PART III—THE WAR YEARS (1939–1945)

Chapter

23 Marriage, War and the Wood-gas Producer 127

24 'Man-powered' in Brisbane 131

25 Mica Mines Revisited 137

26 Out at the Salt Lakes 139

PART IV—POST-WAR YEARS (1945–1951)

Chapter

27 Building Road Trains 141

28 Early Cattle Transport 150

29 The Famous 'Drum Case' 153

30 Acquiring Wings and Crashing Things 161

31 Prospecting at Bonya Hills 165

32 The 'Case' of the Missing Gold 166

33 Lucky Escapes at Rum Jungle 168

34 Being 'Conned' by Con 174

35 Desert Drama in Lasseter's Country 176

36 Solar Power 189

PART V—MORE TRANSPORT AND MINING

Chapter

37 Back to Cattle Transporting 199

38 Mining at Jervois Range 205

39 Trials and Tribulations at Jervois 214

40 Joe Baldissera and the Turquoise 218

41 Return to Jervois 222

42 Bonya and Molyhil Mining 224

43 Drought, Floods and Fires 233

PART VI—RETIREMENT (1979 onwards)

Chapter

44 'Mulga Express' Mark IV and my Second Wood-gas Producer 241

45 Keeping On Keeping On 244

APPENDIX A: Family Tree Chart. 246
APPENDIX B: Map of some of the Pastoral leases in Central Australia, 1914 247
APPENDIX C: Map Showing my Transport Operations in the N.T. 248
                Diagrams: Self-tracking mechanism of my trailers 190
                               Sketch Plan of the Wood-gas producer 243

FURTHER READING 249
ALPHABETICAL INDEX 250

PART I
DEEP WELL (1915–1928)

CHAPTER 1
THE HARDY SOULS OF PIONEERS

Among the red sand hills and desert oaks of Central Australia, on 11th January, 1915, my father, Gerhardt Johannsen, acted as mid-wife in delivering me safely to his wife, Ottilie, in a little log hut where we first lived at Deep Well, 50 miles (80 km) south of Alice Springs, or Stuart township as it was known until 1933, in the Northern Territory. Earlier, in 1914, Dad had started building a bigger four-roomed stone house in which we were living by the time I was christened at the age of four months.

The Reverend Bruce Plowman, an A.I.M. minister who had travelled to Deep Well by camel for the occasion, wrote these words (recorded in his book *The Man from Oodnadatta*) as he rode a camel back from Deep Well Station:

> "On Sunday afternoon the family gathered in the dining room and after a short service little Kurt Gerhardt was solemnly baptised to the joy of his parents...."

Father, Trudy, Mother, baby Kurt and Elsa on Kurt's christening day, 1915

My father, Gerhardt Andreas Johannsen (known to many as "Joe"), was born in Denmark on 14th November, 1876. His father's name was Matthias. Dad left Denmark in 1899, working his passage as a crew member on a ship, intending originally to join his brother Nickolas in South Africa. However, the Boer War (1899 – 1902) altered those plans so he sailed on to Australia, jumping ship at Adelaide, South Australia.

He was a stonemason and builder by trade and plied his skills in the Barossa Valley. He helped build the Saint Thomas Lutheran Church in Stockwell, South Australia in 1904, although the bell tower was added later. More of his building work can be seen at the Hermannsburg Mission, Northern Territory and in the ruins of other buildings in 'The Centre', constructed after leaving Hermannsburg Mission.

He met my mother, Marie Ottilie (Tilly) Hoffmann, in the Barossa Valley. Her parents were Karl Johann Hoffmann (1843 – 1883) and Maria Dorothea Gunster (1851 – 1909). Mother was born at Ebenezer, South Australia, on 15th July, 1882. As a young woman she earned money by working in a cannery at Angaston during the fruit season. This enabled her to pay for dress-making lessons, valuable skills which helped her provide for the family in later years.

Gerhardt Andreas Johannsen and Marie Ottilie (Hoffmann) Johannsen, 1910

My parents married in 1905 and lived in a cottage Dad had bought at Stockwell, in the Barossa Valley. Elsa Margaret Johannsen was born 21st June, 1906. A little son, Curt, born in 1908, died in infancy five months prior to their going north in 1909 when they responded to a call for help at the Hermannsburg Mission in the Northern Territory. They and my oldest sister, Elsa, then aged almost three years, travelled by rail up to Oodnadatta. From there they continued by horse and buggy to the Hermannsburg Mission where they remained for two years.

At Hermannsburg Dad's work included erecting buildings, stockyards and general repairs. He also taught some of the aboriginal men building and other working skills. Mother taught general housework, sewing and handcrafts to the women. Part of the story of her life of coping, courage and strength of character is featured in Jennifer Isaac's book, *Pioneer Women of the Bush and Outback*, Landsdowne Press, 1990, and also in unpublished writings, *Background History and Pioneering History of the Johannsen Family*, by Jose Petrick, Eugene Pfitzner, Trudy Hayes and Kurt Johannsen.

In 1911 Mother, Dad and Elsa left Hermannsburg and went to Deep Well, which was one of a series of stock route wells, probably built around 1880 by Ryan's Camel Party who had built most of the other wells in the area for the South Australian Government from Oodnadatta to Tennant Creek. My second sister, Gertrude (Trudy) Ottilie Johannsen, my best friend and playmate during our childhood years, was born on 28th August, 1912 at Deep Well.

Mother and Elsa travelling on camel, 1910

Deep Well Station

Deep Well was truly named since it was about 218 feet (65 metres) deep. Deep Well Station itself was a square mile of Government reserve land for the stock route well which had first been taken up by the Hayes family in 1884. There were three log cabins with thatched roofs at the well. Dad obtained a Government contract to water travelling stock brought down by drovers from the 'Top End' en route to the Oodnadatta rail-head in South Australia. He later took up the lease of 1,000 square miles of the surrounding country, which was Crown land. (See Pastoral Leases Map in Appendix B).

Camel teams, operated by the Afghans and transporting provisions up to Alice Springs and other properties up North, also stopped at Deep Well for water. The charge was two pence per head for camels and one penny per head for cattle. A camel would drink up to 20 gallons (100 litres) of water per head after a three or four day trek. The cattle and horses only drank half that amount.

Many a time Mum might be found arguing with an old Afghan cameleer who'd say, "Oh no! My camel didn't drink anything much. He only wash his mouth!"

Mum would smartly retort, "Well then, you'll have to pay for washing his mouth!"

The other animals wouldn't drink out of the trough after the camels had been drinking, particularly the horses which were afraid of the smell of camels. When the camels left we'd have to clean the trough out.

The Hayes' property (Maryvale Station) bounded our station on the south, east and north and had been taken up in the late 1800's and early 1900's as part of a grazing licence before my parents arrived up there. They also settled at Owen Springs and Undoolya. Initially, Dad was not very popular taking up a lease of the land right in the centre of their little 'kingdom'. Old William (Bill) Hayes didn't like it at all and tried to stop Father from settling there. (The Bill Hayes family were not related to Trudy's husband, Alan Hayes, whom she later married).

During the 1914–1918 War, with Dad being of Danish and Mother of German descent, things were quite unpleasant for them at times because of prejudice. Dad was referred to as the "German" or the "Hun". However, they endured this and countless other hardships, continuing to work for a better future. They started out at Deep Well, with only twelve cows, one bull and about a dozen goats. They settled in one of the log huts and over the years they gradually built up Deep Well Station.

Some of the stations sank their own wells, but most of them were sunk for the Government by contractors, such as Ned Ryan's Camel Party. The well-sinking parties would set up camp, taking into consideration the availability of water and trees for construction of the wells. The reason why it is often so bare around most of the old stations and homesteads is because 'the nearest tree was the best piece of timber!'

When firewood or fence posts were needed they were carried on shoulders or dragged by a horse, so the trees closest to the homestead were used first. The homesteads were also invariably built close to the best water supply because water had to be carried by buckets to the house. Stock milling around the troughs to drink would turn a place, already denuded of vegetation, into a 'dust bowl'.

Can you imagine the hazards and sheer hard work involved in sinking such a well in the early days? It had to be drilled by hand and blasted out. The easy part was going down as far as the water; the hard part was deepening it to get a better water supply after striking water. The men would start at about five o'clock in the morning, pulling the water by windlass with 8-gallon (36-litre) buckets, as fast as they could go! By about nine or ten o'clock the water would be shallow enough for a man to descend and start digging out the dirt which was blasted the night before while another two people would keep on pulling water out. Then it would have to be drilled and blasted out again. Most of the wells sunk in the early days never had a great water supply because the well sinkers couldn't keep the water out long enough to continue working in them. They also used horses pulling 20-gallon (91-litre) buckets on a two-bucket 'whip' system for the final stages.

A drive would be put in sideways in the well shaft at the bottom, because it was too dangerous for the men to work while the horse-drawn buckets were going up and down. If a worker up top accidentally knocked a chip or a stone in, it would hurtle down like a bullet – Zzzzzzt! If it hit a person on the head or shoulder he'd be done for. The men would often hold a bucket or shovel over their heads to protect themselves when the buckets were being sent up. By putting in a horizontal drive a man could stand in it while the buckets were carried up and down. Often men would work in water waist-deep. When the side drive was in deep enough, say 20 feet (6 metres) it served as a reserve tank, storing the water overnight ready to be used for watering stock the next day.

Father Building our House

Before I was born, Father had set about building a bigger four-roomed house, for which he needed lime. With a tip dray and two horses he drove about 3 miles (5 km) south where there was a lime deposit. He quarried out the lime, loaded it into the tip dray and brought it back to the station. After getting the lime he went out about a mile (or 2 km) to cut logs of wood about 6 feet (1.8 metres) long and 3 or 4 inches (8 or 10 cm) in diameter. He'd cart three loads of wood to one load of lime and stack it near the site where he wanted to dig a pit. The pit would be about 7 feet (2.1 metres) square and about 6 feet (1.8 metres) deep. On one side was a small slit-trench dug down with a hole which only came through as a vent at the bottom of the pit.

He stacked light kindling about 1 foot (30 cm) deep at the bottom of the pit, then added a layer of heavy wood closely stacked and packed about a foot deep. Next he'd add about a cart load of lime, all broken into small pieces, in a layer about 8 inches (20 cm) deep. Layers of lime and wood were alternated until it was stacked up about 3 feet (1 metre) above ground level. It was then set alight from the kindling at the bottom and took about twelve hours to burn through. By then it was a mass of red coals, with the lime also glowing dull red.

When no more flame was visible he laid sheets of iron across to prevent it from cooling off too quickly, covered the edges with dirt and left it for about three days to cool. The lime was then dug out as 'quick lime'. When put into a trough with water the heat from the quick lime caused the water to boil. After it settled down and was stirred, the milk of lime, looking like thick cream, was mixed into sand, making a slurry which was used as mortar to lay stones for the foundations and walls of the house.

Dad gradually built the walls of the house up to the normal height for those days of about 10 feet (3 metres) minimum, which was the way nearly all the early stone and lime buildings were constructed for many generations, right up until about 1925 when cement became readily available for building. Making lime entailed quite a bit of work. Dad took about two years to complete the house, because he couldn't afford to pay much for extra labour and only had one or two aboriginal men helping him.

Most of the timber used for building the house (door jambs, window sills, roofing timber, ceilings and rafters) was hand-dressed desert oak. Dad used 4–6 inch (10–15 cm) diameter logs, specially selected for straight grain. These were split, adzed or broad-axed to smooth them off, then shaped up to whatever was required.

Doors, windows, tables and other items of furniture were made of kurrajong, a softer wood. This was also brought in by the tip dray, with one end slung under the axle and the other end dragging. One log, about 2 feet (60 cm) in diameter and about 10 feet (3 metres) long was stood up with props, then slabbed down, using a cross-cut saw, all done by hand. The men stood on trestles and gradually sawed it down into 4 inch (10 cm) thick flitches. These slabs, in turn, were sawn up into boards and planed for doors and so forth. The floors were paved with sandstone flagstones and the cracks filled in with sifted ashes and fat.

The corrugated iron used for roofing was brought up by camel. The maximum length which could be carried by a camel was 10 feet (3 metres) but mostly the sheets of iron were only 8 feet (2.4 metres) long, loaded in crates. A crate on one side of the camel would be counter-balanced on the other side by another crate filled with bags of sugar, flour and other stores. Camels were always loaded with equal weight on both sides. Especially quiet animals were needed for such loads.

The amount of labour required in the early days for building is mind-boggling – quarrying out the stone, loading it by hand, carting it to the building site using a dray or wheelbarrow, cutting, dressing and splitting the timber and so forth, all with very limited help. The stock yard must also have taken about three years to build, accomplished between other tasks. From studying an old photograph, I estimated there were 110 posts around 10 inches (25 cm) diameter and 10 feet (3 metres) long and 520 rails 12 feet (3.6 metres) long by about 4 or 5 inches (10 cm) diameter. Each post was joggled and bored, and each rail was fitted into the joggles, adzed off, bored and wired. Post holes had to be dug, with many of those in rocky, gravelly ground and gates made. There would have been at least 3,000 man-hours which went into building the stock yards alone! In addition, Dad built the goat yard, wagon shed (50 x 20 feet) and the saddle and harness room (20 x 10 feet). Those jobs alone must have taken four or five years to accomplish.

In the meantime Dad took on various contracts to earn extra money to pay for bought items and food, although we were almost self-supporting with home-produced food. We had goats for meat, milk and butter and later on some cows, which provided milk, cream, butter and cheese. The vegetable garden was carefully tended with water pulled up in buckets by the whip method. A horse was used at first with the whip, and the Dodge 4 in later years until we acquired a pump. It gives truth to the saying "Life wasn't meant to be easy!"

Deep Well Station, around 1924

Chapter 2
Living Near the Desert Oaks

Family Life and Work

The desert oak is a beautiful tree which I have always loved. Besides being used in building our home we collected the nuts from the branches of the desert oak trees which Dad cut. After drying them we loved to eat the nuts; we also made little toys with the 'acorns' or cones.

Every person in the family was important and from a young age we all learned to carry out tasks which were essential for the well-being of us all. My early memories revolve around the daily events of working and playing. Besides pulling water for travelling stock, Dad also worked on contracts such as sinking wells and constructing police buildings at Alice Well and Arltunga and the prison cells at Alice Springs (now preserved as a National Trust building). Dad and Billy Liddle, an early contractor and pastoralist, worked together on many of these sites.

All of our building materials, food and miscellaneous supplies were brought up by camel teams, usually once every six months, with small parcels and letters arriving with a monthly camel mail service. Two of the old Afghan camel drivers, Fred Khan, and his brother Hussein were well known to the pioneer families in the 'Red Centre'. Fred, who was a very active man, lived well up into his eighties, and had changed his name to Fred 'Smith'. (He died in 1982). We always waited eagerly for the arrival of another cameleer, Sideek, hoping he would bring us a few lollies, one of the rare occasions when we ate sweets.

A typical camel train around 1920

There were times of terror in our young lives. My sister Elsa (now deceased) told the story of how Mother became lost, as recounted by her daughter-in-law Jose Petrick in our family history writings. Late one afternoon in 1916, Mother went out to bring the milking cows home. Since there were no fences the cows often strayed long distances away, especially after rains. Dad was away working near Christmas Dam, about 12 miles (20 km) east on the boundary between our property of Deep Well and the Hayes' property of Maryvale. As night fell Mother hadn't come home with the cows. I was only a baby at the time and don't recall any of it, but Elsa, then aged 10, and Trudy aged 4, were very frightened at being left alone after it became dark.

Dad had a premonition that something was wrong and rode his horse back home. It was too dark by then to follow Mother's tracks so he hung a lantern on the highest part of the hill near the house, hoping she would see it and find her way back. She apparently did see the light but thought it was an aboriginal camp fire and may have been afraid to approach it alone and at night. She was also disoriented and confused and instead of walking north towards home, she walked south. This kind of thing can happen quite easily when walking up and down sandhills, especially in the dark. There was no moon and stars are not much help once a person feels lost! Eventually she came to a track which by morning brought her to Breaden's Dam, one of our neighbour's dams, about 10 miles south of Deep Well. Some stockmen were camped there and one can imagine their surprise to see a solitary woman appearing out of the 'Bush'. One of the men brought her home on horseback, much to the relief of our little family.

Not long after that incident Dad was involved in an accident out at Mount Burrell where he was re-timbering the well. One side of the wall suddenly collapsed onto the staging platform down in the well where Dad was working. The weight of the collapsed portion ripped the supporting cable out of the head frame, letting the stage and the whole lot hurtle down the well. Dad just managed to step off the platform in time and stood on the side where there was some new timber. While he had been working down in the well there were two men assisting him up top – a white man and a half-caste lad. The white man 'shot through' from fright, while the young half-caste lad held tightly onto the windlass which was still standing. He wasn't strong enough to pull Dad up so Dad climbed 20 feet up the rope!

In the early days about three families of aborigines camped near Deep Well. Dad mainly employed Harry, Maggie and their son Johnny and his young wife, Lady, to help with work around Deep Well. They were good, reliable people and helped operate the 'whip' drawing water for the travelling stock before the well was equipped with pumps. Alice Perkins (later Costello) lived with Burk Perkins, a half-caste man who sometimes assisted Dad with some of his contract work, while Alice was invaluable in helping Mother with housework, minding us children, helping build the goat yards and so forth. (Alice, in her nineties, is still living in the Old Timers' Home, Alice Springs in 1992).

After getting up in the morning, the daily routine for us children included 'pulling the bed together' and having breakfast, usually ground-wheat porridge boiled up with a bit of salt, with sugar and milk on it. We didn't very often eat the more expensive rolled oats or oatmeal. Wheat was bought for the 'chooks' as well as for our own consumption, and one of our jobs was to grind the wheat up with a little hand coffee grinder.

Trudy picking golden wattle and wearing a typical square-cut working garment

Kurt with his straight pants held up by braces

Each morning the girls milked the cows and goats and let them out to forage for food. They also separated the milk, made butter and helped Mother with various tasks. The job of bread-making was shared among all of us once we were old enough to help, with Mum or Elsa usually doing the baking in a Metters wood stove located in the closed-in back verandah which served as our kitchen. There was no such thing as refrigeration for keeping food in those days. For our meat and other perishable food we used a Coolgardie safe with wet hessian bag hanging around the sides and top.

Mother's dress-making skills stood her in good stead, as she made all our clothing. The work-play garments were only simple, square-cut things, but serviceable. My pants were just straight up and down, held up by braces. Mother could and did make some beautiful garments, but for us children and the aborigines' working gear she just made durable outfits using calico or striped galatea. A dress was virtually straight up and down like a nightgown with sleeves in the corners and a hole in the middle for the head! Some of Mother's fine sewing is on display in the Old Timer's Museum at Alice Springs.

A Close Call

I almost killed myself when I was about four years old. While Elsa was in the next room setting the table I climbed up onto a stool which was beside the big cupboard. I knew all the goodies for making cakes and biscuits were kept up on the top shelf, so I grabbed the first bottle I could see. I took the cork off a bottle and gulped it down, choked and fell off the stool. It turned out I had swallowed pure Essence of Lemon, which was about 90 per cent alcohol.

Elsa had previously heard Mother advising some women on what to do in case of poisoning and other emergencies, so she grabbed a jug of milk from the table and poured it into my mouth. I vomited up a lot of the essence, but I had also inhaled some of it into my lungs. I was unconscious for a couple of days and succumbed to pneumonia. I must have been delirious because I can recall weird images and nightmarish thoughts, from which I couldn't escape, going on in my mind. My parents didn't know whether I would pull through or not for several weeks.

As an aftermath of that escapade I was very weak for a long time and inclined to be a bit on the 'weedy' side as a child, tiring very quickly. I remember one instance when Dad wanted some charcoal for the blacksmith shop and took us three children over to the camel camp, about a quarter of a mile away. That seemed like miles and miles for me to walk! We were given one lolly for each billycan of charcoal we brought back. The other children gathered about four billycans full of charcoal while I only managed to get about half a billycan. They'd have two billycans full before I even managed to walk over there!

Christmas time

At Christmas time I remember Mum, Elsa and Trudy making dozens of biscuits, using cut-outs which Dad had fashioned in the shapes of men, ladies, ducks and so forth. I helped by icing the biscuits and thoroughly enjoyed that task. Around 1921 we had a very wet, cool Christmas with very high humidity. The dampness caused the biscuits to soften and break off at the stems where we had tied them to the Christmas tree. I recall seeing the beautiful iced biscuits falling down out of the tree. We were allowed to eat some of them if they fell off, but Mum rescued a lot more and dried them out in the oven.

Dad made a beautiful rocking horse out of a kurrajong tree trunk with one of the limbs making the neck. He must have spent weeks on it, making it as a Christmas present for Trudy and me. It was painted mottled grey, with a proper face, tail, mane and stirrups and it rocked perfectly. The most expensive rocking horse that one could buy today, made of wood, would not equal the standard of this one! It lasted for many years and even when we were all grown up Elsa's children had it out at 'Mount Swan'. Eventually it fell apart and was used as a swing, hung on a couple of ropes from a tree.

We also enjoyed taking it in turns tinkering around in the blacksmith shop. If one of us was making presents for birthdays or Christmas no one else was allowed to come in there! Elsa also was very good at making exceptionally pretty Christmas cards using cardboard and pressed flowers, gluing them on and painting the surrounds.

More Family Life

My parents were very strict in the traditional European manner of those days. I don't particularly remember them showing me much physical affection, although I know they loved me. In those days a boy was generally treated more harshly than a girl if there was any wrong-doing or disobedience. Dad made a special whip from an old whip handle about 18 inches (45 cm) long with a double leather strap of heavy harness leather, about half an inch (2 cm) wide, also about 18 inches long. This hung in an accessible place ready to be used when discipline was called for. From about the age of six I remember that Trudy would get a 'telling off' whereas I would receive a whipping around the legs or shoulders, which I thought was very unfair. I had an affectionate nature and wanted to be loved, but often it seemed I couldn't do anything right to please my parents. This gave me a man-sized inferiority complex for a long time. I was also very shy, as was Trudy (and probably many other youngsters born in the bush) because of living in such isolation from other families.

Life was harsh for the animals as well as people in those days. Dad went out hunting one day, taking his kangaroo dog with him. He walked up into the hills about a mile away from the house and sat down on a ledge to rest for a while. The dog lay down near him, just below the ledge. A giant perentie (a very large monitor-lizard), which apparently had been hiding under the ledge, flew out and attacked Dad's dog, killing it. Dad shot the perentie which measured just over 10 feet (3 metres) long. Some of the larger perentie back in the early years were so big even the aborigines would steer clear of them. There are very few of the giant ones existing today, if at all, as most of them have been shot out.

Not long after that incident Dad arrived home, after being down south, with a beautiful little collie pup for us children and another kangaroo dog for himself. I really loved that dog which we named Laddie. I cuddled him and let him smooch all over me, giving and receiving some of the affection I craved. About four years later Laddie was killed by dingoes which used to sneak up around the goat yard or come after the chooks. Laddie and Dad's kangaroo dog generally chased after the dingoes and helped to keep them away. However, earlier Dad's kangaroo dog had died from a poison bait which had been laid out for the dingoes. On the night Laddie died we heard spooky noises of howling and yowling during the night, then everything was quiet. Next morning Laddie was missing, so we went looking for him. Although Laddie had killed dingoes in the past, this time he was out-numbered; we found only the carcase of our dear dog. I felt really sad about losing my true friend.

Elsa, Kurt and Trudy with Laddie, around 1919

Dad had quite progressive ideas and implemented many of them at Deep Well Station. Around 1921 he put up a tank near the windmill and piped water into our house. He also made a bathtub and installed a chip water heater and shower in the bathroom, which was really something in those days. After that he added a netted-in sleep-out to the house which was great to sleep in away from the buzzing mosquitoes and flies, as well as being cooler.

One thing the netting didn't stop was a centipede which wriggled into bed with me and bit me about six or seven times! It started its journey on my shoulder and crawled down inside my pyjamas onto my stomach and legs. Each time it bit me I'd grab at it until at the finish there was no more poison left in the thing! I must have kicked up an awful fuss because everybody raced out and lit lanterns. For quite a while they couldn't figure out what had bitten me. When they saw the first bite marks up on my shoulder they thought it might have been a snake, but eventually they discovered the centipede and killed it. Those bites made me quite sick and caused a high temperature. In those days the usual remedy for anything was a dose of castor oil – real castor oil which tasted awful! It would kill any poison in your system, if it didn't kill you first!

Around 1921 Mother started teaching Trudy and me grade one reading and writing. We had all the necessary books and Mum would sit us down and make us work at our lessons. Later, shortly before we went to Hermannsburg in 1922, the Correspondence School was started and once a month the South Australian Education Department sent up our sets of schoolwork with the camel mail. That made our lessons quite a bit easier for Mother to supervise and more interesting for us with coloured pencils for colouring and so forth. I completed grade one and had started on grade two. Generally, however, I hated school lessons – I think I was dyslexic. The strap was always there, hanging on a nail ready to be used for any wrong-doing or not getting on with my work (I was a great day-dreamer!) It must have been hard for Mother struggling to organize lessons in addition to all the daily chores and running the station, especially when Dad was away on contracts.

Trudy, Kurt and Elsa reading

Play

If we worked well at our lessons for a couple of hours and had finished our daily chores, we were allowed to go and play down by the trough or the creek. Often we would go down to a sandbank in the creek and jump off. In childhood many things seem much larger to us than later in life – a shrinking seems to happen! That sandbank seemed very high then but the highest point from which we jumped was only about 8 feet (2.4 metres). The bank was mostly red sand with a bit of clay in it, enough to make it solid, so we could dig into it and carve out cubby holes.

Trudy and I were very sure-footed, probably from practice chasing each other around the stock yards on the top rails and down to the trough, which also had a rail over the top. We jumped from one stump to another, on up to the top of the 10,000 gallon (45,000 litre) tank. We'd run around the edge of that and across the pipes, bare-footed (and while Mum and Dad weren't watching, or they'd have had a fit!) I seldom wore shoes until we moved to Alice Springs when I was about fourteen.

We also loved to climb up into the gum trees to pick off 'sugar', created by little insects called Prillya, from the leaves. We used a half-inch (1 cm) rod about 15 feet (4.5 metres) long, with a hook on the end of it and climbed up it, pinching the rod between a big toe and second toe, pulling up with our hands. Then we'd hang with one arm looped around the bar, picking the 'sugar' off the leaves.

For Mother and my sisters the main pleasures and pastimes were sewing and embroidery. Elsa, who was the bookworm of the family, also really loved reading. She always had her nose stuck in a book, which often got her into trouble. We also played chess and I became quite good at the game because I had the ability to calculate many moves in my mind and remember them. Mum and Dad were pretty good players also. One time, around 1924, a 'RESO' party came up from Adelaide, bringing their chess boards with them. Although I was quite shy, when I heard somebody mention playing chess I volunteered, "Oh, I'll give you a game." The chap who had suggested playing didn't want to have a game with me, but the other men urged him, "Oh, come on, give the kid a game!" He reluctantly agreed, "Oh, all right." He probably thought he was going to wipe me out in half a dozen moves. After the third game he received a lot of 'ribbing' from the others because I had wiped him out three times! (Apparently this chap was a bit of a pain and a braggart). That was quite a feather in my cap and helped boost my self-confidence.

Another night, after we children had been sent to bed in the sleep-out, our parents were playing cards with some visitors in the dining room. We'd turned off the lantern and were settling down to sleep when we heard a strange rustling noise on the floor. I struck a match and saw a large green snake slithering under my bed. I screamed, which brought Mum hurrying out with the .410 shotgun at the ready. She shot the snake right under my bed while I nearly flew up to the ceiling from the noise of it! Mum was a pretty good shot when she needed to use a gun!

One frosty winter morning, when I was about six years old, Dad walked into the house with a surprise for us. He carried in some big slabs of ice which had formed in the cattle trough. It was the first time I remember seeing ice. We raced outside and down to the trough, bare-footed over the frosty ground and lifted slabs of ice out, chewing it up and having a really good time playing with it.

My first trip to Alice Springs in a buggy with Dad and Elsa was quite exciting for me. We left home at dawn and completed the 50 mile trip by dusk. We had four horses in hand and I recall travelling through Emily Plains, about 10 miles of flat, open country where the horses really went mad racing towards Heavitree Gap. By the time we reached the Gap and struggled through the sand the horses were exhausted. It took us quite a while to get them through there. Once inside the Gap on the 2 mile drive into town we passed through salt bush which grew as high as the buggy. We could hardly see over it – it was a terrific sight!

The Australian Inland Mission (A.I.M.) had just started on the foundations for building Adelaide House. We camped at the home of old Charlie Sadadeen, an Afghan gardener. He had fig trees growing and many kinds of vegetables. Charlie showed me how to eat the figs so I wouldn't get a sore mouth and said I could have as many as I wanted. The next day I had the 'trots'. He had forgotten to tell me about that!

Kurt holding Father's horse, Ned

A Corroboree

The only real corroboree I ever attended was staged about a mile from Deep Well, around 1921. Our family went out after dark in the buggy and watched. The aborigines wore huge, elaborate decorations on their heads which I thought were marvellous. It was like going to a concert for us. We were allowed to stay until about nine o'clock when the real or 'poison' corroboree as the tribal elders called it, took place for tribal men only, and all the women and children had to leave. Dad took many flashlight photographs of the event which unfortunately were lost in a fire which later burned my sister Mona's home in Alice Springs.

Not long after the corroboree there was a related tribal killing among the aborigines. Apparently one of the aboriginal women had been found watching a part of the corroboree which was taboo for women. Her punishment was death. The elders secretly chose someone to 'point the bone', the identity of whom was only supposed to be known by one or two of the elders. On this occasion the information had leaked to others in the tribe and they found out who did it. A big fight was triggered off, because her death had to be avenged by her family members, resulting in many spear and knife wounds.

Revenge killings are rarely heard of these days, but in earlier days the warriors preferred to maim another man rather than kill him, thus causing him to be dependent on the women to feed him and look after him. I can remember Mum doctoring up one warrior with bad thigh wounds, about 3 inches (7 cm) wide. The only attention his wound had received was having ashes rubbed into it to dry it up, but it had become infected. Such a person would walk again eventually, but only with a stiff-legged gait. The aborigines believed that kind of punishment was far better than death because the man was forever 'hamstrung'. It was demeaning for him to have to be hand-fed by the women – the worst thing which could happen for a warrior. Besides, if they killed another man, the revenge killings would continue. Wounding a man brought the 'pay-back' to a neutral position.

Wells and Water

Also around that same year, Dad was working out near Christmas Dam, putting down a bore, using the old hand-boring plant. This comprised a spring pole or sapling 20 feet (6 metres) long, with the largest end clamped into a tree about 10 feet (3 metres) above the ground, with the centre propped up on a post the same height. There was a pulley on each end of the pole and a windlass or winch anchored to the tree. A wire cable from the winch passed over the two pulleys and down the drill hole onto the drill bar, which was rammed up and down by four men pulling on the four ropes attached to the end of the pole. This method was successfully used for drilling holes to a depth of 150 feet if the operators didn't die from exhaustion before they reached that depth!

As children, another of our jobs was to help pulling water. We usually had an aboriginal worker on top of the well and another walking the horses up and down, pulling the long rope which drew the water up in two 20-gallon (90 litre) buckets – one down and one up. The long rope, called the 'Whip' (about 280 feet or 85 metres long) went around a large pulley along the flat. As the horses were led in one direction, one bucket would go up and the other went down. When the full bucket was landed and emptied into a 'landing trough' it was swung off to return down the well again while the other full one would come back up.

Later, Dad put our first pump in the well – a small pump jack and an Eclipse petrol engine, only one and a half horse-power, which was far too small for the purpose and gave lots of trouble. In 1922 Dad obtained a Horwood-Bagshaw kerosene engine which was about three and a half horse-power, to replace the little Eclipse. This did a better job and made us one of the best-equipped stations in 'The Centre'.

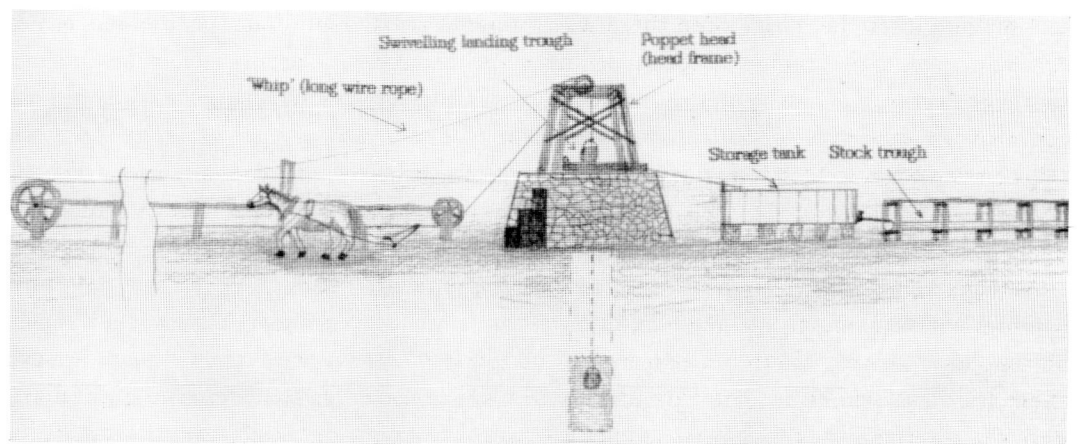

Diagram showing 'Two-bucket Whip' method of drawing water from the well.

The little Eclipse motor was transferred into the blacksmith shop, previously the little log hut where I was born. The hut had been converted into a blacksmith shop and storeroom. The Eclipse motor drove a shaft which Dad made and connected by belt to the old type wooden-barrel washing machine. It was also used to drive a small saw for sawing bush timber into planks for doors and so forth.

The Horwood Bagshaw motor, which was a kerosene engine that started on petrol, was often hard to start. Many a time when Dad was away on well contracting I'd be cranking away on it trying to get it started. I'd wind and wind, using the heavy cast-iron handle which was attached to a short piece of the crankshaft. If the motor fired when I was tired out, the handle would slip off and WHACK! it would hit me in the side of the face. I still wear a couple of scars above my left eye from that!

About once a week the old-type leather belt would gradually stretch and become loose. We boiled up some sticky 'goo' made out of resin and fat, similar in consistency to treacle and applied it to the belt to make it grip better. When the belt was too far gone we'd stop the engine, cut a small section out of it, rivet it together and 'make do' with it again.

Diseased Cattle

Around 1922 a mob of about 800 head of cattle were being brought through from the 'Top End' from around Newcastle Waters, a 'pleuro' infected district. Not much was known about pleuro-pneumonia in those days and the drover bringing them down hadn't realized his mob was infected until some animals started dying. By that time he'd passed through a dozen cattle stations. When infected cattle drank out of a trough the saliva bearing the germs went into the water and any other cattle drinking from that trough would contract the disease.

It caused a big panic when it was realized that this mob of cattle were diseased. In those days vets were not available up there so people had to make their own serum. The formula was known to Sergeant Stott in Alice Springs. Blood from an infected beast was bottled, mixed with water and kept warm to let the 'wogs' incubate and breed up. The clear serum was then collected and after it settled out they'd dip cottonwool thread in the serum and poke it into the tails of the animals, treating the whole herd. If an animal contracted pleuro-pneumonia the effects would go into the tail where the antibodies built up. They'd become pretty sick and usually the tail dropped off. We only had a herd of about 400 cattle at the time, and after Dad had treated all of our animals we had about 100 of them with stumpy tails, but at least it saved them. Some station owners who didn't do anything about the epidemic lost large numbers of cattle.

Also in 1922 our telephone was re-connected. During World War I (1914–1918) it had been removed because Dad was of Danish-German descent. It was a bit ridiculous because the range of the telephone was only about 100 miles at that time and my parents were hardly likely to be spies out in the desert! However, those were the rules of the time. Having the telephone back on also saved a lot of cattle for other people because it was possible to phone through to the next station, relay a message to them which enabled them to be prepared for the infected herd coming through. That herd, which distributed the disease, reached Oodnadatta with less than half their herd left. It was a huge loss to the owners, besides leaving a trail of disease on the way which caused further losses of several thousand cattle.

I also recall the terrible bushfires following the very wet season in 1921. Dry grass and undergrowth was prolific and bushfires were sweeping the countryside, putting everyone into a panic. Every available person had to work hard, bashing fires out and back-burning to try and keep them away from the station. After about the third day a shower of rain helped put the fires out, which saved our whole property from being burned out.

We didn't realize it then, but 1922 was also the start of the devastating drought which lasted for seven years, bringing untold hardship to many people living in 'The Centre'.

One Sunday morning Martin went over to Kramer's home and was appalled to find him shaving on the Sabbath! Martin was so disgusted he gave him a great lecture, calling him a hypocrite for professing to be a man of God. Martin never spoke to Kramer again after that!

On one of our trips into Alice Springs we stayed with their family. The Kramer's daughter, Mary, and I decided we were going to be "very best friends". We were both aged about ten years at the time. I was putting my best foot forward, showing off what I could do, and so was she. To prove to her that I was really strong and had a hard head, I started banging my head on the wall which had a doorway with no frame in it. Pastor Kramer really went 'crook' at me for being silly and trying to crack his lime and sand bricks!

After we returned to Deep Well I always wished to go back again and continue being friends. We did see each other a couple more times on visits to Alice Springs or when they came to Deep Well, but Mary and her brother Colin went south to attend school. They had attended the first school in Alice Springs with Mrs Ida Standley, who started the school for the half-caste and some white children. Later Mrs Standley moved to Jay Creek with the half-caste children. Standley Chasm was later named after her in recognition of her work, since she and the children discovered it. I didn't see the Kramers for many years until the early 1930's when Mary came back up North. My sister and some others were trying to do a bit of match-making between Mary and me, but nothing came of it. I didn't meet her again until about 1985. She had never married and was living on her own.

Part of the Afghan hawker's camel team and his aboriginal helpers

Chapter 3
Visitors and Bush Folk

The White 'Lubra'
Trudy and I were down at the cattle trough watering the cattle one day around 1920, when visitors travelling by camel buggy arrived. We saw a white woman, her husband and some aborigines. The woman, dressed in a long dress of blue and white striped galatea, elastic-sided boots and wearing a felt hat, came walking towards us. We raced up to the house to tell Mum, calling out excitedly, "Quick Mum! There's a white lubra coming!" (We hadn't seen many white women before). The visitors were Arthur and Mrs Bailes from Bloods Creek Station, near Oodnadatta.

They remained at our station for a few days because Mrs Bailes hadn't enjoyed the company of another white woman for some time. The way they mumbled when talking fascinated me. As I furtively stared at them I finally realized they had no teeth. I wondered how they could possibly eat meat without any teeth. My curiosity about that matter ended during lunch, when Mrs Bailes pulled a mincer out of her calico tuckerbag and clamped it firmly onto the table between herself and her husband. They took the meat off their plates and put it through the mincer!

Mother, at that time, spoke very little English, but apparently the two women got along very well in spite of the language barrier. At home we all spoke German since Mother had been born and reared in the Barossa Valley and attended a German school and German was Dad's second language; English was our family's second language at home and in dealing with other people.

'Behold He Cometh Quickly' Slowly!
Travelling ministers were other visitors passing through Deep Well. Around 1921 the Reverend Kingsley (Skipper) Partridge came through by camel. This is mentioned in the book *Camel Train and Aeroplane (The Story of Skipper Partridge)* by Arch Grant, Rigby 1981.

In 1923 an evangelical Baptist gospel missionary, Pastor Ernest Kramer, his wife Euphemia (Effie) and his family also visited. Elsa, Trudy and I were playing up on a sandhill, a short distance south of the house, when we saw a covered wagon approaching in the distance. It was pulled by camels and had two goats tethered behind it. The Kramers were on their way up from south, intending to settle in Alice Springs. With the slow-moving caravan they only travelled about 10 miles a day. We thought this was a great joke because on both sides of the caravan Pastor Kramer had written in large, bold writing, "Behold! He cometh quickly!"

Their home in Alice Springs, made of lime, sand bricks and bush timber, would have been about the eighth or ninth home built there. Church services were held in a bough shed which Pastor Kramer built as a house of worship for aborigines to attend. Pastor Kramer was a very devout man and lived a good Christian life. However, in 1924, when government houses were being built, a chap by the name of Herb Martin, a building contractor, came up to work and became quite friendly with the Kramer family and, being of the same religion, often visited them.

The Afghan hawker

One visit we always enjoyed was the Afghan hawker with his team of about ten camels. On each side of every camel there was strapped a big box about the size of a low-boy, covered with canvas to make it water-proof. These boxes were filled with everything imaginable, from trinkets to shoes, boots, hats, shirts, dresses, blouses, patent medicines, scented soaps, cheap perfumes and sweets. You name it, he had it! His arrival once a year was like Christmas to us. The perfumes were very popular with the aboriginal women. Instead of paying wages, Mum and Dad would make up a parcel of goodies for them at Christmas, including clothes (which they selected themselves), a blanket each, a towel, boots, two large red handkerchiefs, a comb, a packet of needles and scissors, two reels of thread, soap and scent, a cherrywood pipe (which wouldn't break like the old clay pipes), a leather belt, a butcher's knife, a tomahawk, an 8 x 10 foot (2 x 3 m) ground sheet (which they called a 'tent'), plus a quart pot, a pannikin and a billy, plus lollies and clothing for the children.

Friendships

When I was about seven years old some visitors arrived from Alice Springs, in a buggy. It was the Price family, including Alf, Ronny, their sisters Pearl and Molly and their parents. They stayed for several days before returning home. Being terribly shy, I was so nervous when they first arrived I hid behind posts and trees, just looking at them. Eventually we broke the ice and started to play. Pearl and I liked each other and decided to be friends, but our good will towards each other was short-lived when I started playing a game with one of her brothers. Pearl wanted to be in it too, but her brother insisted she wasn't allowed to be in the game. She got the sulks and said she was going back to Alice Springs, so off she walked down the road to the lumpy ghost gum, about half a mile away. She sat there on the rocks on her own for about an hour, got rid of her sulks and came back again. About fifty years later when we met up again, reminiscing about old times Pearl and I had a good laugh about that incident!

I seldom had other playmates except for Trudy and Emily, the daughter of Alice Perkins, the half-caste woman who acted as a Nanny and helped bring me up from when I was a toddler. Alice was Mum and Dad's very loyal general helper, housemaid and child-minder.

A Famous Visitor

I saw an aeroplane for the first time in 1921 when Francis Birtles flew over on his way to Darwin. He was a well-known aviator, writer and explorer. From then on I looked at aeroplane pictures in books and had a hankering to learn to fly one day. Before that trip he had come through by car and stayed at Deep Well for several weeks. On that trip he was surveying places to land his aeroplane for refuelling on his forthcoming flight and waiting for some petrol to be delivered by camel. He'd sent some up a month earlier, but it was contained in square kerosene tins and the constant rubbing on the camels fatigued the metal, causing it to leak out. When it arrived there was only about one-third of the consignment left. The next consignment was sent up in 8-gallon drums, which all arrived safely.

Francis Birtles with our family on a stop-over at Deep Well, 1924

Later on, in 1924, Birtles came through again, driving from north to south in a Bean car. He had it rigged up with about six spare tyres stacked on a bit of piping on the back. I remember that quite clearly, although they didn't stay long that time. They just refuelled and were off again.

1922 was the last time I recall a line party coming through by horse-drawn wagon. The next party came through by truck in about 1926 or 1927. The early line parties comprised a wagon with about ten horses, the wagon being laden with telegraph poles, insulators, wire and camping gear. Some of the men rode the horses. Their job was to repair any telegraph poles which were falling over, damaged or rusted, and clear away scrub growing up underneath or near the wires. These trips generally took a couple of years to travel one way from say Marree or wherever they started from, right through to Darwin, then later they'd work their way back again.

The rumour was that by the time they returned, after two or three years had passed, they'd be meeting up with the offspring which they'd created on the previous trip. Like the sailors, some of them had a reputation for having 'a girl in every port'. Quite a number of the half-caste population in the Territory possibly originated from some of these workers in the early days. Some of the men later married their aboriginal women and settled down with their families in different parts of the North.

More Visitors and Motor Cars

In 1924 the motor mail run was instituted from Oodnadatta to Alice Springs and Arltunga, taking over from the camels. Sam Irvine was the first driver of the motor mail. Gradually the camels were phased out except for general loading. Even later on trucks were used for that. Wallis Stores, later Wallis Fogarty's, were originally storekeepers and general agents in Oodnadatta and then Alice Springs.

They used two Dodge G-boy one-ton trucks and a one and a half tonner, employing Billy McCoy, Phil Windle and Claude Golder as some of their first drivers. Later on some Reo and Willys Knight trucks were brought in for carting loads of materials for building the new government residences and the Courthouse. At first the vehicles had a lot of trouble getting through the sand, but gradually the Government sent parties down to 'corduroy' many of the sandy river crossings. Aboriginal prisoners were also used as labour for some of those works.

People could also hire prisoners for private purposes for a couple of days at a time if they needed a cellar dug or something similar. The old warder, Jim Shannon, who was about 70 years old and fairly frail, walked around with his .44 Winchester rifle over his shoulder, which I think was more for appearance than any useful purpose, because if the prisoners had really wanted to attack him they could easily have over-powered him and taken his rifle.

Gene French, Harry Wolf and Edgar Horwood in an old 1923 model Reo mail truck at Deep Well. Trudy is standing on the right.

Some other visitors who arrived 'out of the blue' around 1924 were a team of surveyors coming through surveying for a proposed railway line from Kingoonya, South Australia to Alice Springs. This party arrived with four T-model Fords. They were having problems with punctures in the old beaded-edge type tyres. When one tyre had too many holes they'd stretch that tyre over the top of another slightly smaller one, giving them a double tyre thickness, preventing some of their 'stake' problems.

My life-time interest in mechanical things, especially motors and vehicles, was sparked at a very early age when some of these visitors to Deep Well came and needed repairs to their cars and buggy wheels and so forth, and of course by my father's aptitude for blacksmithing and his interest in similar things.

Chapter 4
Trips South

I was too young to remember our family's first trip to Adelaide and Stockwell, South Australia in 1918. Mother, Elsa, Trudy and I stayed in Father's little cottage in Stockwell in the Barossa Valley for nine months near to our relatives, while Father obtained a position as a guide, driver, cook and general help to the eminent geologists Sir Baldwin Spencer and Dr Keith Ward. They arranged for Dad to guide them on a buggy expedition through 'The Centre' and part of the Simpson Desert. Mr Ochtman, a telegraph operator who was on sick leave due to tuberculosis, was left in charge of Deep Well Station; he wanted to stay in a warm climate for the sake of his health. Dr Herbert Basedow was another eminent scientist whom Dad took out in his horse-drawn buggy to various locations in the Territory.

Our second trip south was in the winter of 1922. Dad was taking his first mob of about 100 fat bullocks down to the Oodnadatta rail-head from where they would be loaded onto the train for the market in Adelaide. Previously Dad had sent a few cattle away by including them in a mob from another station. 1922 was our peak, after the good rains of 1921, when Deep Well Station was going nicely and our stock starting to breed up. Actually, it was just before the downturn of the long drought from 1922 to 1928 which nearly wiped out all the cattle, but we weren't to know that then!

Mick Kerin (a cousin of Joe Costello who worked for Dad), Elsa and some aboriginal men helped Dad in droving the cattle down to Oodnadatta. They all took turns doing a two-hour night watch each night. It was a long, slow journey in the horse and buggy keeping pace with the cattle; it took us about four weeks to reach Oodnadatta.

At dawn everybody rose and the cattle were moved on. After breakfast the buggy was re-loaded and followed on behind the cattle, eventually passing them so that we could set up for the dinner and evening camps and the cooking was also done for the next day. About once a week a beast was slaughtered for meat, some of which was salted and some kept fresh for about three days. The rest was given away to station people as we passed through in exchange for watering our cattle and horses. Occasionally some would be given to passers-by who were short of meat.

We were approaching Horseshoe Bend, with the horses laboriously pulling the buggy along the Finke River for about a mile in very heavy, white sand. The poor horses were exhausted and needed to be spelled about every half mile. Horseshoe Bend Hotel and Cattle Station was about 80 miles (128 km) south of Deep Well, run by Mr and Mrs Augustus (Gus) Elliott and their daughter, Sheila.

I tasted lemonade for the first time at Horseshoe Bend. When we arrived the men went into the pub where I could hear quite a lot of talking going on. I wondered what was happening so I walked in and found Dad standing by the bar. He picked me up and sat me up on the counter. One of the chaps said to Dad, "Come on, buy him a lemonade!" I didn't know what that meant but I was given a 'lady's waist' glass of crystal-clear, fizzy liquid. I felt terribly awkward and shy in

Elsa and Mrs Ruby Elliott at Horseshoe Bend Station, 1922

front of everyone and nervously fumbled and fiddled around with the glass. Dad started to 'rouse' at me to pick it up and next thing I'd knocked it over and spilt it. I grabbed for the glass and quickly gulped the last spoonful which had survived the spill. I thought it was the loveliest stuff I'd ever tasted. Of course, I wanted some more and the men were going to buy me another glassful, but Dad, being embarrassed, sent me out saying I was "useless and clumsy." So that was my first taste of lemonade – one spoonful!

I also clearly remember the night we camped after leaving Horseshoe Bend. I woke in the middle of the night with a terrible crawling, gnawing sensation in my lower bowel. I was terrified and quite certain that I was fly-blown and would surely die! (I had seen some fly-blown goats back at home and they had died). I was too scared to say anything to Mum or Dad, but Mum noticed how restless I was and finally found out what was troubling me. She pronounced that I had pin-worms for which, of course, the cure was the inevitable dose of castor oil! I didn't know which was worse – the worms or the griping pains from the castor oil, but I didn't die!

When we came to the Finke River near Old Crown Station it was in flood from heavy rains a week earlier in the MacDonnell Ranges. Luckily we hadn't camped the previous night in the river bed as there had been no sign of rain at that point. The cattle had to be coaxed into swimming across the muddy waters with a rider going ahead to guide them across. When the cattle safely reached the other side the buggy was unhooked from the half-floating harnessed horses as they couldn't pull it through the water. The horses were taken over to the opposite bank, then the buggy was hitched by a long rope to the horses' harness with us children remaining seated in the buggy as it crossed the swirling waters, wetting some of our luggage.

The old stationmaster's cottage at Coward Springs, South of Oodnadatta

I also recall that when we camped for a couple of days at Charlotte Waters, which was a telegraph repeater station near the Northern Territory/South Australian border, Dad and Mick made a little boat out of a piece of bark, complete with a tiny sail. It floated across the water which seemed to my child's eyes that it stretched a long way across, but was probably more like about 50 feet (16 metres). Also, a couple of the old aborigines, Noornee and his friend, fashioned ornaments such as pipes, birds and kangaroos, out of the high-quality kaolin (clay) found near Charlotte Waters.

The next place I recall was when we camped at the Angle Pole waterhole which was about 6 miles (10 km) north of Oodnadatta, while waiting for the train. I was attracted to some little bushes about a foot high which I had never seen before, which smelled like peppermint. I tasted them and sure enough, they tasted just like the little black peppermint lollies we occasionally ate. Strangely enough, about 1960, when I went down to Oodnadatta carting cattle in my road trains for old Jim "Ironbark" Davey from Granite Downs, I revisited this waterhole and found the peppermint weed was still growing around the edges.

I remember seeing an old steam traction engine at Oodnadatta, behind Wallis' store. It had been bought by Steve Adams who was related to Bill Hayes, grandfather of young Teddy and Strat Hayes. They had intended to drive it from Oodnadatta to Maryvale Station but only got as far as Angle Pole where the boiler burnt out. It was towed back to Oodnadatta by a team of horses and deposited behind the store where it still stood 50 years later. Nowadays there is no store and the old engine which had been rusting away out on the flat has been taken to a museum. Many places were removed when the railway line was extended from Oodnadatta to Alice Springs. Oodnadatta almost completely died and became only a siding, but it has revived and survives as a small town with less than 200 people (in the 1980's). Over the past decade the people of Oodnadatta have developed a commendable sense of civic pride which was formerly lacking.

However, in 1922 Oodnadatta was the biggest town I remembered seeing, since I was too young to recall anything of the previous trip to Adelaide and Stockwell. There were two separate sections of the town then alongside each other – the Afghan town and the general town. The Afghans were the main transporters with their huge camel teams in those days. I think there were about 300 Afghans living there at the time. Many camel teams operated from Oodnadatta up to Alice Springs and places further north. The camel teams also travelled east and west of Oodnadatta to the remote desert places and homesteads. Marree, further south, was also a big camel team base.

Our family boarded 'The Ghan' pulled by a steam engine for the rest of the journey down to Adelaide. I remember going clickety-clack across the Algebuckina Bridge, which was about a mile long, the longest bridge in Australia in that era.

Algebuckina bridge

Adelaide and Stockwell

During our stay in Adelaide we stayed in the 'Grand Coffee Palace' in Hindley Street next to West's Theatre (which is still standing now, known as the Plaza Private Hotel). I was given a penny ice-cream made of pure cream and firmly told I could have only one – it would make me sick if I had more than one!

From Adelaide we travelled to Stockwell to visit our relatives again. Trudy and I attended school for about six months. We stayed in Dad's little cottage again, although Elsa stayed with Pastor Stolz and family of Light Pass so that she could be prepared for confirmation.

In the meantime Dad was looking for work to earn some extra money, as the cattle had only netted about £1 a head after freight and agent expenses. Having previously taken Sir Baldwin Spencer and Dr Keith Ward on a trip around the Territory, he was recommended to explorer Valhjalmur Stefansson, a Swedish anthropologist, who wanted to study the lives of aborigines living in the wild in the Simpson Desert of Central Australia.

The cottage where we stayed in Stockwell

Stefansson wanted to compare how the aborigines survived in the extreme dry heat of Central Australia with the way Eskimoes survived the extreme Arctic cold. He had studied and lived with the Eskimos for twelve months. Stefansson later gave Dad an autographed copy of his book describing his experiences living with the Eskimo people. When I was a lad I enjoyed looking through that book with its many pictures, letting my imagination run wild thinking about what it might be like living in those conditions.

Dad earned £100 a month for the six month expedition with Stefansson. On his return he surprised us by arriving home in a 1922 model Dodge 4 vehicle which was the first privately-owned motor vehicle in Central Australia. All the electrical equipment was removed from it and Dad had it converted into a utility truck with the front seat widened on both sides to seat the five of us. It cost £312 brand-new, less £12 for not having electrical parts. It had a buggy-type hood with no curtains or side screens, and a little box on the back for luggage. The car was equipped with 32 x 4-inch tyres on 24-inch wooden wheels. Since there were no mud-guards over the front wheels you can imagine what happened driving it in the rain! The mud flew up and spattered into our faces, but did that dampen our spirits? Never! We just held a wheat bag over us and drank in the great thrill of travelling in an automobile, waving to the people we passed travelling in buggies with their startled, 'spooked' horses. With Father just learning to drive there were some exciting moments, as with most learners. It was all go or stop, particularly in rough or slippery places.

Return to Deep Well

For the return trip to Deep Well we proudly set off from Stockwell in the Dodge 4, staying at Clare the first night and Melrose the second night. Mum called it "Smellrose" because the maids at the hotel where we stayed apparently had not cleaned the rooms properly. In those days the toilets were away down the back of the hotel near the stables, so the rooms were equipped with chamber pots under the beds.

The road from Stockwell to Quorn hardly qualified for the name since it was really only a track, which made for a rough journey. At Quorn the car was loaded onto the train as far as Oodnadatta. From there we drove on home via Bloods Creek Station, crossing several wide sandy creeks and on to Charlotte Waters Repeater Station. Any who have travelled the Oodnadatta Track in modern times can imagine what it must have been like back in the early 1920's. We passed through Old Crown Station, Horseshoe Bend and the Finke River, continuing on through the Depot Sandhills (about 30 or 40 feet high) which are part of the western edge of the Simpson Desert. The track crossed the Hugh River about seven times in 11 miles (17 km) before we reached Alice Well Police Station (which Father had built earlier). Maryvale Station was our last stop before reaching Deep Well.

Father with our first car, a Dodge 4 utility, stuck in the sandhills

Luck was in Father's way for our trip back in the car from Oodnadatta to Deep Well, because there had been rain a few days earlier. Most of the sandy areas were firm enough for the 4-inch high pressure tyres which cars used in those days. Dad had also had some sand grips made to go over the back tyres. These grips were two pieces of leather belting with hoops all the way round to make the tyres wider for negotiating sandy creeks and sandhills.

By the time we arrived home at Deep Well Dad almost had the hang of driving! In those days there weren't any driving schools – you just hopped in and did your best.

Undoolya Homestead, showing typical style of early station homesteads, with Dad's 1922 Dodge 4 utility on the left and an identical one ordered by Ted Hayes Senior at Undoolya

Chapter 5
Interlude at Hermannsburg 1922 – 1924

At the end of 1922, soon after we returned from our visit south, the Lutheran Church needed help and asked Dad to return to Hermannsburg Mission Station as manager while they were looking for another missionary and new staff. Pastor Carl Strehlow had died suddenly in October, 1922. Dad and Mother accepted the call and arranged for Joe Costello to look after Deep Well Station while we went to Hermannsburg. My sister, Mona Dora Johannsen, was born at the Mission on October 27th, 1923, soon after we arrived there.

I found living in Hermannsburg was very different from living at Deep Well. Mother and Dad tried to introduce new ideas to the people there, such as making and eating whole-meal bread, which our family always ate. The aborigines didn't like that at all. They reckoned it was no good – it was "chook food" because it was made from wheat and ground on the spot. They were going to go on strike rather than eat that and were backed up in this attitude by the young relieving missionary, who agreed with them.

Another idea which Mother suggested was for the women to shorten their dresses. The old-style dresses were quite unsuitable for living conditions out there as they dragged along the ground in the dust.

Mother and baby Mona, 10 days old

New issues of clothing were made up with the hems about a foot (30 cm) from the ground. This also caused a great protest. The old missionary had always given them dresses which touched the ground, even for the children. There were many other ways in which our parents tried to improve the living conditions for the aborigines. They met opposition along the way but eventually most of them came around and accepted many of the changes.

Mother, with Elsa helping, taught sewing, embroidery and crotcheting to the aboriginal women so that they could make their own clothing. Also during this time, once the fortnightly camel mail run came out there, we received more of the early correspondence lessons from the South Australian Education Department, which we'd started at Deep Well. Mum and Dad, much to my disgust, put Trudy and me to work on the lessons. We did approximately twelve months of lessons while living at Hermannsburg. Trudy completed grade 5 and received a book prize for her writing and I completed grade 2 level.

Dad began a tannery at the Mission, which Pastor Friedrich Albrecht later continued, teaching the aborigines the tanning process and how to make their own boots. As well Dad taught them some woodwork skills, making mulga wood ornaments and that sort of thing.

A vegetable garden was established at Kaporilja Springs. Dad suggested piping water 3 miles (5 km) into the Mission but that idea was opposed by the Mission Board. It was not until twelve years later, in 1935, long after we were gone that Pastor Albrecht finally organized for it to be done, still against the wishes of the Board. He raised the necessary £3,000 through donations, collected by Una and Violet Teague from Victoria, and the pipes were supplied by James Hardy & Company, Melbourne at cost, which made it possible to go ahead with the project. The pipes were laid and a large storage tank was built from local stone and lime. Pastor Albrecht was also the first Lutheran pastor to introduce training of aboriginal pastors.

The Mission aboriginal men were supplied with 12-gauge shotguns and ammunition. They had all been instructed to save the empty shells and return them so that they could be re-charged at the Mission. In those days, possum skins were valuable, although the skins in the north were not as good quality as those from a colder climate. I remember seeing four or five wool bales full of possum and rabbit skins being sent away by camel. After the drought was in full swing the skins were no longer any good.

In 1923, early in the drought, Dad organized the building of stock yards around the waterholes near the Hermannsburg Mission. There were thousands of brumbies out there, eating out the countryside, depleting the feed supplies for cattle, and also causing a shortage of water. The only way their numbers could be controlled was to get them into trap yards and cull them by shooting. The outward trap gate was left open so that the brumbies became accustomed to going in and out of the yards. The trap was then sprung and next morning the slaughter would begin. About 2,000 brumbies and donkeys were shot around that time.

Brumbies near the stockyards at Hermannsburg Mission, 1923

Cutting timber for the stockyards at Hermannsburg, around 1910

During that drought I remember seeing rabbits jumping out of the trees. They'd clamber up to 10 feet high (3 metres), trying to get a bit of food from the leaves of the mulga trees. Sometimes we'd see dead rabbits jammed in the fork of a tree. There were even kangaroos hanging in the trees where they had struggled up to reach a bit of green or moisture and become trapped. The bodies of thousands of kangaroos, rabbits, cattle and horses lay scattered all over the ground where they had died near the dried up water holes.

One day, while the trapping was taking place, a really violent thunderstorm came through. The stockmen were making camp and the horses, which had just been brought in, were milling around restlessly. They must have sensed that something was wrong. Suddenly a hailstorm hit, with hailstones as big as hens' eggs hurtling from the sky. The men grabbed their saddles to protect their heads while the horses bolted. The storm belt was only about a mile (1.6 km) wide but when we drove out there afterwards the hailstones were stacked up deep in the gutters. Some of the men and several of the horses were badly bruised.

The shooting of the wild horses used to upset me. The .44 cartridges which were used in the Winchester rifles were re-charged and I helped do that. When I discovered what they were being used for I decided I was going to save some of those lovely young horses from being shot. I took a whole packet of the new caps destined to be put in the cartridges. Before they were put in I dropped them into water and then dried them off so that no-one would know.

A day or two later, when there were about a hundred horses in the trap, the men were lined up ready to fire at the brumbies when the gates were opened. They always let the horses run out as they shot them in the ribs so that the horses would run about half a mile before they dropped. This saved the men from having to drag the bodies away from the water, as the working horses were too poor in condition to use for that hard work. Of course, this time the rifles didn't fire. I received a helluva beating with a doubled up stockwhip for that misdeed! Somebody must have seen me tampering with the cartridges and 'squealed' on me.

During our time out there, Dad and about twenty aborigines made a track down to Palm Valley for several government parties who were surveying for potential development in the Northern Territory. There was the Dutton party in 1923-24 and later the Murray Aunger and RESO parties. Most of them were driving Dort cars, built in Adelaide, South Australia, the first cars built in Australia.

In 1924 Pastor John Riedel arrived in Hermannsburg on a temporary appointment for six months and other staff also arrived, so our family returned to Deep Well.

The Murray Aunger party with the Dort cars at Hermannsburg. Note the sand grips on the back wheel.

Note: The RESO expedition included members of a service organization or Old Boys Club from Melbourne and Sydney called the "Resorians," who were in the Northern Territory in August, 1927 – photographic records are held in the N.T. Collection of the State Library of the Northern Territory.

Chapter 6
Hard Times Back at Deep Well

In 1924 we moved back to Deep Well to face some very bleak years. The terrible seven-year drought lasted for the rest of our time at Deep Well and was devastating to our family, especially when illness struck my parents. Our cattle died by the hundreds, leaving very few young cattle to brand. One of our biggest and constant jobs was dragging dead cattle away from near the trough. The starved, weakened cattle came in to have a good drink and lay down to rest near the trough. Then they'd be too weak to get up and would die on the spot.

Daily routines continued with Dad seeking extra work to provide for the family while we children assumed more and greater responsibilities as we grew old enough to handle a variety of work around the Station. Dad acquired several government well contracts, renovating some old wells and sinking new ones, which often kept him away from home. He bought another vehicle – a Dodge one-ton G Boy truck for the job so that we now had two vehicles. Joe Costello, Mick Kerin, Trott and Sonny Kunoth, and Musty Milnes also worked with Dad on these contracts which kept us going through the drought. The income from watering travelling stock, other than a few camels, had ceased.

Around 1925 Dad sank a well on Woodford Creek near Pine Hill Station. The station couldn't afford to pay him cash, but they had sheep so they reimbursed him with 100 sheep. If the drought had broken all would have been well, but having to walk the sheep about 150 miles (240 km) carting water for them over the long, dry stages to Deep Well was quite a job. After they arrived at Deep Well Station it was another burden of work to shepherd them, trying to find very scarce food and water and also to guard them from the dingoes.

Two more children were added to the family during these hard years – Randle Werner Johannsen born 10th August, 1925 and Myrtle Edna Johannsen born on 22nd November, 1926.

Early Inventiveness

I think the idea of the self-tracking trailers for my road trains, which I built after World War II, originated in my imagination when I was about ten years old. I loved to make all sorts of weird contraptions out of tin lids, kerosene boxes and other bits and pieces and hooked my models together with two, three and sometimes four trailers with wire towbars around bent nails. I pulled them around with a piece of string or a stick. I made a prime mover with a steering front axle and used a long stick to push and steer it around by twisting the stick.

I also made up a wind-driven prime mover using a 15 inch (36 cm) fan with twelve blades, cut from a piece of galvanized iron. I soldered it onto a shaft with a small pulley running down to a seven-pound treacle tin with route tracks on either side and a piece of tube rubber for a belt. This model worked quite well on a side wind, but would only pull two trailers. I abandoned that project because it wasn't strong enough to pull any trailers.

During 1925 I would sometimes disappear into the blacksmith shop building a model steam engine, which was quite a success. I never got around to making it mobile, so it remained a stationary steam engine. I fashioned the boiler out of an old float chamber from a cattle trough and used an inverted valve from a car tube as a safety valve. The crankshaft was made from a quarter-inch rod bent into both the cam and the crank. The cylinder head was cast out of some zinc and I used nails for valves and the springs out of old car valves. I'd heat up the boiler on the forge and demonstrate it for visitors. I didn't think very much about it because it was just another of the things I loved to fiddle with, but everyone else seemed to think it was quite marvellous when I actually got it going. We kept it for several years until later, after we moved to Alice Springs, I was demonstrating it one day when the crankshaft broke. I threw it in the corner and didn't think any more about it. It probably ended up down at the tip, but I am sorry now I didn't keep it. In recent years I've promised my family I would build a replica for them to keep as a souvenir of my mechanical achievements in childhood.

Too much to Drink

One day Dad, Joe and Musty Milnes went to Alice Springs in the old Dodge, taking me along with them. Dad dropped the two men off at the Stuart Arms Hotel and said, "I'll pick you up here again about an hour's time." Joe went off to buy some shirts and other items, while Musty went off to the hotel. After about two hours Dad returned and found Musty 'out to it' lying on the footpath, talking to himself, outside of the hotel. He was 'blind drunk'. Dad hadn't known that Musty had a drinking problem – he couldn't handle much grog and would get drunk on the smell of a rum bottle cork! It was a surprise to us because Musty was an excellent musician and played the violin beautifully.

In those days when there was no 'canned' music around, different people would gather together to provide their own entertainment. Often at home there'd be Musty playing his violin, someone else playing a mouth organ, Dad with his flute, Mum playing our old portable organ and other people singing. We also owned a cylinder phonograph, but people preferred to make their own music.

Early Radio

In 1924 John Flynn and Alfred Traeger arrived up North with the first radio receiver and transmitter. We installed a tall aerial up on the hill and the two men stayed at Deep Well for a few days testing it and using morse code to get contact with 5CL in Adelaide. The radio only came on intermittently with a programme of music, ten minutes of static, followed by news, more static and more music. The old sets could only be used at night with much squealing and howling trying to tune them in. However, that was the beginning of the development of pedal radio and the Flying Doctor, which proved to be such a marvellous help for the pioneers of the Outback. The first radio link in 'The Centre' was established about a year later when Flynn and Traeger came back with a pedal radio transceiver linking Hermannsburg with the base at Adelaide House in Alice Springs.

During dust storms the aerials picked up a lot of static electricity. We children played with the sparks which arced off the end of the wire, up to an inch and a half or 4 cm long, holding up a stick and watching the sparks rattle out onto a water pipe. We burned patterns into paper and so forth, and although we were careful it is a wonder one of us didn't get an electric shock or 'boot'!

More Growing Up

Growing up is hard at times. A child must learn the rules of work, play and gradually assume responsibilities. In the process we suffer pleasures mixed with pain and we come to realize that things are not always fair or easy.

Dad kept a cocky (galah) on a stand and chain. This cocky would chew and chew at his chain to try and break it and escape. One day he managed to break a link and flew around the homestead really enjoying himself. Guess what happened? When Dad came home and caught it, he put it back on the chain and Cocky got a belting with the whip (the one I dreaded with the two straps) and I also got a whipping because Dad reckoned Cocky couldn't get off the chain on his own, so I must have let him off!

Another time I was talking to Cocky and scratching the back of his head, when Cocky suddenly decided to sink his beak into my finger! Naturally, I screamed and yelled with pain and at the sight of blood dripping from my finger. Out came the whip again. Cocky received another hiding for biting me and I got a hiding for going near him to be bitten. You couldn't win!

Many a time when I'd received a thrashing from Dad or Mum with the stockwhip or a piece of siphon hose, I'd make up my mind that as soon as I could walk properly again and the red welts had faded (although they never ever left any lasting marks), 'tomorrow' I'd start stashing away some food and when I had everything ready I'd run away! However, I'd start wondering where could I run away to? Alice Springs was 50 miles away and if I went there it would soon be reported and I'd be carted back home and given another belting. A couple of times when I threatened aloud to run away Dad would say, "Go on, then. Run away! But if you come back here again or if we have to go and get you, you'll get another belting!" Anyhow, I never did run away and thousands of children before and after me have made threats of running away from home and been treated in similar fashion by their parents. With all the frustrations in life up there we all had to be tough to survive.

From the time I was about nine years old, another of my jobs was to kill a goat and butcher it for meat when needed. I'd catch a goat in the yard and ride it out, steering it by its horns, to the gallows up on the hill near the yard. Most of the goats were too strong for me to man-handle at that age. Occasionally I had to ask Trudy or Elsa to help me if it was a very big animal. I had to cut its throat and put it on the gallows before skinning it and carting it down home to hang up. Elsa would cut it up the next morning.

Although we were brought up on the station and often saw goats and other animals being born, I never gave a thought to where human babies came from. One afternoon, however, I was coming down from the goat yard, past the old log hut behind which the aborigines camped, carrying half a goat which I'd slaughtered. One of the women hurried over to me and said, "Get the missus! Mary bin have a baby!" The baby was lying in the red dirt and I suddenly noticed that it still had its umbilical cord attached. It was a shock as I instantly realized that babies must be born in the same way as the animals. In their natural state the aborigines never had anything with which to wash a new-born baby so they just lay them in the dust which was later brushed off – and they survived.

There were always chores to be taken care of in the garden and chopping wood, cleaning the ashes from the kitchen stove and taking them down to the pit toilet, and cutting up newspapers into squares for use in the toilet. We took it in turns with wood chopping. Some of the aboriginal men would also chop wood and we'd have a bit extra on the wood pile. After Dad purchased the Dodge truck in 1924 we used that to cart wood for the homestead.

Elsa and Mona were allergic to "itchy grubs" but they didn't affect Trudy or me. One of my jobs, when the caterpillar nests were seen up in the big ghost gums, was to climb up, pull them down and burn them. One time I was after a nest way up top of a tree which I couldn't quite reach so I got a hook and pulled the nest down. In the process the dust and grubs from the nest showered all over me. After I had collected all the grubs and burned them I went home to clean up and shower. I went into the bathroom, removed all my clothes, shook them out, cleaned up and put them back on again before going off to do something else. About half an hour after I'd been in the bathroom, Mona, who was about four years old at the time, went in to wash her hands. She'd only been in there about a minute but in less than an hour she started running a high temperature (104 degrees Fahrenheit) and was one big red blotch all over from contact with the dust from the itchy grubs still floating around in there. It is amazing the severe effect that dust has on susceptible people.

The New Bike
When I was about ten years old a mail order Sampson's catalogue was sent to us, which listed all sorts of gear, amongst which was a bicycle costing £8. There was also a cheaper model costing £4. One of the young aboriginal men, Burk Perkins, wanted this pushbike so Dad ordered it for him. Burk had saved up his money at the rate of five shillings a week, the amount he received in cash wages over and above his keep until he had enough for the £4 bike plus an extra £1 for freight.

When Trudy and I saw the new bike and tried it out we wanted one too. After proving to Dad that we could ride it in the sandy conditions, and pointing out how useful it would be for riding the 3 miles (5 km) down to New Well to water the cattle, he ordered the £8 one for Trudy and me. I always wished it was the man's model which was much nicer as far as I was concerned, but Dad ordered the lady's model because Trudy and I had to share. It certainly came in very handy during the drought when we took turns riding to New Well to water the cattle every second day.

There were thousands of bindi-eyes and many, many punctures so I'd put a spoonful of pepper and 3 or 4 spoonfuls of water in the tube. Whenever a bindi-eye went in and came out a little grain of pepper would be sucked into the hole and the water would seal it up with the gooey mess which was floating around inside the bike tube. That handy little hint I have passed on to many lads and lasses with bicycles, right up to the present. It works better than some of the commercial substances because it doesn't mess up the whole tube. Also a piece of wire or cord across the front and back forks, almost touching the tyres, will flick the bindi-eyes off before they penetrate right into the tyres.

After my tenth birthday, another of my jobs was to grease the windmill about once a fortnight, which was necessary because of the old-fashioned open gearing. One day I climbed up and found the wheel spinning around on the shaft itself – the key had sheared off. It was quite a big job drilling it out with an old type ratchet drill with a chain over the hub. It took about a day to drill it, after which I hammered in a half-inch (12 mm) bolt, which lasted for quite some time.

When that eventually gave way I went up to put in another key. A sudden gust of wind came up and rotated the wheel around with me in it! It was quite exciting at the time, but not the sort of thing you'd want to do too often. I was lucky it was only a small gust of wind and thus escaped unscathed. I had forgotten to tie the wheel to the frame before I climbed into it since it was calm at first. I thought it was okay – or rather at that age, I probably didn't think about it at all!

Trudy and I sometimes stood up on the mill's platform, looking down. We reckoned that if we had a big umbrella, like the beach umbrellas we'd seen in the catalogue, we could jump and sail down using it as a parachute. I'm glad we didn't have a beach umbrella, otherwise I might not have been here to tell the tale!

Dad's Illness

My childhood virtually ended with Father's illness, which came in the middle of the drought when the sandhills had encroached right up to the front verandah of the house. The picket fence which Dad had built around the lawn was buried under sand and we could walk straight over the gate or the picket fence, or anywhere at all!

It was towards the end of 1926 when several members of the family became ill with high temperatures and pain in their limbs, but Dad was quite badly affected. It turned out the illness was poliomyelitis, or 'infantile paralysis' as it was often called then. There were several other cases of it in and around Alice Springs among the European population. Dad became paralysed, which was a great hardship for him and the whole family. Mother read up about the illness in her old German homoeopathic medical books and followed the instructions, alternately wrapping Dad's limbs in sheets wrung out in hot and cold water hourly. This must have been an extremely difficult time for Mother, as such procedures were quite time-consuming and all regular chores still needed attention, to say nothing of the demands of her other young children and baby Myrtle.

So that Dad could have the quiet and rest he needed, Trudy and I had to take our little sisters, Mona and Myrtle (aged 4 years and 1 year at the time) and brother Randle (2 years) away from the house, armed with refreshments, rugs and cushions, and keep them amused down in the sandy creek bed.

Elsa had come back home after Dad became ill. She had been sent away in 1925 when she became pregnant and her fiance, Martin Kleinig, who was very ill with tuberculosis, had sadly committed suicide early that year. Those were the harsh customs of the times, but she was needed by Mother and returned home in April of 1926 with her seven-month old son, Martyn. She helped look after Mona, Randle, her own little son and new baby sister, Myrtle.

After a few months, when he was well enough to be moved, it was decided Dad needed further treatment in Adelaide. Sam Irvine, the mail contractor, offered to drive Dad in his truck to Oodnadatta. He made up a bed and canopy for Dad on the back of his old Reo truck and looked after Dad all the way to the rail-head at Oodnadatta where he was transferred onto the train down to Adelaide under the care of one of the Australian Inland Mission (A.I.M.) sisters. Dad was admitted to the Royal Adelaide Hospital for several months. The doctors there apparently praised Mother's excellent nursing and care of him back at Deep Well.

This time must have been a great strain and quite distressing for Mother. Trudy was only 13, I was 11 and we only had Alice Perkins, our Nanny, and a couple of aboriginal families to help with running the station, as the three youngest children were too young to do anything. All the stock had to be watered, the well kept in mechanical repair and the daily chores kept up. Mother became very run-down during the time Dad was away.

Mother with a brood mare

I was needed to look after the pumps, engines and all the mechanical jobs. Looking back now, I would be horrified to see a child of my age then doing what I did, such as shinnying down the well to repair a pipe, 50 or 60 feet (20 metres) down, with another 150 feet (50 metres) below me to the water level. I had to put a clamp on the pipe where it had worn through. However, in times of duress, such actions had to be taken and it would generally have to be Mother or me. Several times Mum went down in a rig, something like a bosun's chair – a board with two ropes on it with a quarter-inch wire rope on the windlass. She sat in it while we children carefully let her down to put new leathers or washers in the pump and then we pulled her up again. At a depth of 210 feet you couldn't call out and be heard up top. We devised a code – one tap for stop, three for up and two for down, knocking on the pipe. If the weather was calm and the person down the well had a very strong voice, it might just be heard up top.

It has been said that "Necessity is the Mother of Invention". The ability to improvise and makeshift is a most valuable asset for a person who has to survive in the bush. When ordering materials or spare parts in the early days there was a minimum of three months elapsing before we could expect to receive them. An order would be sent away for say a tyre or an axle or a set of spark plugs. The order would go down by camel and it might reach Adelaide, the nearest centre, about four weeks after it was put in the mail. If everything went according to schedule the order would probably be packed and ready to go to Oodnadatta by train, possibly another two weeks. Then, when the next camel team was due, possibly six weeks later, it would finally arrive at Deep Well, and if one was very lucky it would be the right part.

Under those circumstances, the art of improvisation became a natural habit. Dad and I didn't even think about a replacement part at first. We looked at the problem and decided whether another one could be made or the existing one repaired. Could something be forged up in the blacksmith shop? Or could I use something else? Or could I make a temporary repair? That was a natural part of life for me. An instance of improvising was when the plunger in the well pump broke off right at the bottom, so a new piece had to be forged up and riveted together. There was no welder, no lathe and very little to work with, barring the forge and a hand-operated drilling machine and rivets.

One time, before Dad became ill, an axle broke in the Dodge. The spare one was also broken, so we had a look at the axle. The drive end was a square end which fitted into the differential. Dad stoked up the forge with a heap of charcoal. Using a big anvil and a large sledge-hammer, he set to work swedging it out, gradually lengthening it until it was the right length and then hammered a square onto the end of it, which worked until a new axle arrived about four or five months later.

The Stolen Meat

An incident happened while Dad was still at home, ill with polio, before he went to Adelaide for treatment. It was my turn to go to New Well to check the water for the cattle. The horses were too poor in condition to ride and petrol was too expensive to use the Dodge so I had to ride the bike. It was hard work because of the sandhills and sand drifts on the road. I detoured around them by following cattle pads off the road, running along mulga flats and approaching the well from a different direction.

The Government had let a contract to two well sinkers to deepen the drive in New Well to improve the water supply. We supplied the men with meat from our station when we killed an animal for our own use. This time I was taking some dry salted meat, tied to the handlebars of the bike, down to them. McIntyre, one of the contractors, was a big, rough, unpleasant, ginger-haired bloke. I was only about twelve years old and very timid. I handed McIntyre the meat and told him it was all we had until we could find another 'killer' beast, or else we could get him a goat. He let go with a string of abuse at me and told me he didn't want "any of that shit" and threw the meat onto some timber out in the sun.

I was always rather scared of him and his mate with their teasing and threats, so I hopped on the bike and bolted for home after watering the cattle and checking to see if there were any dead cattle to be dragged away from near the well. If so, I would need to come back in the Dodge ute. Dad had taught us to drive the car from the time we had first bought it, so Trudy and I were quite competent little drivers by then.

On my way back home I had to pass the contractors' camp by about 200 metres. As I drew level with their camp on the down-wind side I smelled something like a steak grilling, but it didn't register at first. Riding along about a mile further on I saw a flock of crows back in the scrub and a few eagles. I went over to investigate and found a very large and fat freshly-killed bullock, with about 50 pounds (23 kg) of meat removed and the remaining 500 pounds (230 kg) just left to rot. (Sometimes, during the drought a fat bullock, which had been out eating parakeelya – a type of pigweed – would wander in from the desert. These animals could survive on the parakeelya indefinitely without water if there was enough of it around, just as desert camels do. Those beasts which wandered in were the only source of meat we had during the drought).

McIntyre and his mate had thrown some bushes over the carcase to camouflage it. If they hadn't done that I may not have taken much notice of it, because there were many cattle perishing from effects of the drought. As I passed it the penny suddenly dropped – the smell of the cooking steak, the salt meat they had left in the sun and the big bullock with the meat hacked out. It all added up, along with their cheeky threats and insults. A cold shiver ran up my spine.

I rushed back to the station and told Dad, who was in bed. To verify that it was the contractors who had killed the bullock, Dad sent one of our aboriginal workers back to check on the tracks which he found led back to the contractors' camp. He also found two empty shells near the carcase which the police eventually matched up with McIntyre's rifle. Dad told Mum to ring the policeman, Constable John Clow (Jack) Mackay, at Alice Well. Constable Mackay only had horses which were in very poor condition, so he arranged to meet Mum and me halfway on the 60 miles (100 km) journey. I drove the Dodge as Mum couldn't drive.

The two men were arrested and the following day Mackay loaded his two prisoners onto one of Wallis Fogarty's trucks going to Alice Springs. They were charged with cattle killing and having the meat in their possession. They were prosecuted by Sergeant Stott, tried by Special Magistrate Ben Walkington and sentenced to six months hard labour, plus costs of £25, plus £5 for the bullock, which was top market price in those days, to be paid to Dad. It wasn't much for a bullock when the costs of droving to Oodnadatta, rail freight to Adelaide, agents' charges and levees were deducted.

A couple of weeks later Trudy and I rode down to the well to do some mustering and see what feed was around for the animals. While we were walking through the scrub a big stake of mulga wood plunged into the calf of my leg. At the time I was wearing short pants and didn't have leggings. My leg subsequently became badly infected so I was unable to ride the bike for about three months, leaving Trudy to do the run down to the well during that time.

Up in the hills, about 8 miles from the station there was a rockhole which had filled with some water after a thunderstorm had gone through. There was also a bit of feed in the sandhills, but the track into that rockhole was extremely rough, really only a walking track. To make a better track so we could bring the stock to the water, I brought some explosives up from the station to blast away some of the big ledges. Our aboriginal family helpers, Harry, Maggie, their son Johnny and his wife Lady, helped me do that. These people were the most faithful helpers to our family one could ever wish for. Making that track enabled us to bring the camels in to water and load up with water to take back to the camp. We walked the sheep up there every second day for about a month or more until the water dried up and the sheep had to be moved on again.

We tried to keep our few cattle from perishing, but the only water was at Kangaroo Well and New Deep Well. The poor animals were so skinny they could hardly walk. Some of the bigger bullocks would march out up to 15 miles away to feed, staying out and coming back every second night. We had to make sure the troughs were kept full so that the animals didn't have to wait for water. Even so, many perished out in the bush, having walked out too far.

I was out at Kangaroo Well one day, about 15 miles (24 km) away, repairing a windmill pump. I was about 12 years old at the time. We had some aborigines camped out there watering a few of our remaining cattle. I drove the old Dodge, running it on kerosene because it was cheaper and we only used petrol for starting it.

While I was working on the pump I could see a thunderstorm brewing out east. I decided to send Johnny Baboon with two camels back to the station for more stores and some petrol for starting the Dodge. It took a couple of days before we could even find the camels. Eventually, when they came in for water, Johnny caught them and went off next morning, back to the homestead. Back at the station Elsa and Trudy had a couple of jobs for him to do so they asked him to camp there for the night and bring the stores back to us next morning. When Johnny went out early the next day to look for the camels he discovered they'd broken their hobbles and walked all the way back out, almost to the spot he'd started from the day before! That meant another day passed before he could return to the station, load up and come back with our stores.

In the meantime, back at our camp, the aboriginal women had found the track of a big carpet snake. They tracked it all day through the sand and spinifex and found it had entered a rabbit hole. They started digging it out. One of them came back to the camp to get a shovel so I decided to go back the couple of miles to help them. A bright moon shone over our digging activities which took us until about ten o'clock that night before we extracted the snake. It was huge! We triumphantly carried it back to the camp where old Harry immediately stoked up the fire to cook it. I went to sleep, but at about two o'clock in the morning they woke me to have a feed. Everyone was starving hungry. I was offered the liver, which was prized as the tid-bit – the best part! It was about 3 feet (1 metre) long, and narrow. The snake itself was about 10 feet (3 metres) long and as thick as an arm. However, I didn't like liver at any time, so I declined and settled for some of the delicious white meat.

The following morning Teddy Hayes from Undoolya Station came through with a plant of horses, heading back towards his home. He was nearly out of stores too, but he had a couple of johnny cakes (small dampers) which he left with us, so we ate snake and damper for lunch. That evening Johnny returned with the camels and our stores.

A few days later the boys saw some kangaroo tracks heading east to where there had been a shower of rain a few miles away from us at a place called Kooraroong, the legendary place where the Dreaming Time giant kangaroo and emu tracks are imbedded in the mudstone. During the time I was out there they told me the legend of how a giant snake had chased the animals and made a big gully in the hills where those tracks were imbedded in the stone, which I was shown. They were about 10 miles east of Kangaroo Well in the side of a hill with the rock sloping down to the south. I think they were actually dinosaur tracks imbedded in the sandstone.

There was a large flood-out area where kangaroos usually congregated. We needed fresh meat, so I got the Dodge going and headed out there with the aborigines and their kangaroo dogs, which Dad had brought back from Adelaide. I shot one kangaroo with the .22 rifle and Johnny caught two kangaroos with the help of the dogs, so we had plenty of meat. The women had also dug out some yams from the banks of the creek.

One problem – I couldn't start the Dodge because there was no petrol left with which to prime it. There was only kerosene in the tank. I usually carried a spare bottle of petrol but it had been broken. Those old Dodge cars had a fixed crank handle and not to be beaten, I took the spark plugs out so the crank handle would be easier to turn. I put it in reverse and Johnny and I took it in turns winding it backwards up a steady rise. We smoothed the track off and wound and wound for about an hour and a half, and eventually got it about 100 metres up this slope. I thought if we could get a good run up, and put some boiling water in the engine to heat it, it should start on kerosene. Everything was set, the spark plugs were put back in again, everybody on board and we let her go! We rolled down at a great rate of knots to the bottom of the hill, but not a sound from the engine. For some reason I had switched off the magneto and had forgotten to turn it on again. We gave it away for that evening as the old bus didn't have any lights.

Johnny dug a hole about 3 feet long, 1 foot deep by 1 foot wide and made a big fire in it. When the fire had burnt down a bit, one of the roos was buried in the ashes. First it had been gutted, its fur singed off and the tail removed. The tail, roasted on a separate fire, was ready for eating in about one hour. We ate damper, yams and roo tail washed down with a billy of tea for our supper. The kangaroo itself took about four hours to cook. That night we slept near the fire and kept stoking it all night, as it was very frosty. After a good breakfast of roo, damper and tea, to give us strength to wind the Dodge up the hill again, we finally got it going. These were some of the trials and tribulations I learned to cope with. I think such experiences taught me a lot, not only to have patience, but to 'look before you leap'. It all helped me greatly later in life.

Early interest in Minerals

The beginning of my interest in geology and how the earth and rocks were formed began when Trudy and I roamed the nearby hills exploring. We found molten sandstone and many little caves where we could see how the original sandstone had 'boiled up' in places. Johnny Baboon also told me about some very black, "heavy stuff" way out along the hills.

Some time later, I took the Dodge out to draw water at New Deep Well, hooking the car onto the Whip. When I had finished I decided to go and have a look at the black rock he'd been talking about. Johnny and I drove out through the sandhills and scrub and found the place he had described. We broke a piece of the rock in halves and saw glistening, shining streaks through it. I thought it might be silver and became quite excited about it. We headed home and had only driven about a mile when we ran over something sharp which sliced one of the brand-new tyres, only put on the day before. We'd been waiting for months to get those new tyres! Here I was with this 3-inch slit which would need to be bolted together like all the other old ones. When our tyres split we used a piece of old tyre, trimmed the edges and bolted it into the damaged tyre, using about six gutter bolts along each side of the slit and a piece of tube rubber on the inside to stop the tyre from chafing. This generally repaired them quite well.

There must have been a 'hoodoo' on me for that trip because we had only gone about another mile or so when the oil pump stopped pumping. The oil pipe had broken and we still had about 10 miles to go. The old Dodge 4 was splash feed which meant the oil pumped up into the trays where the bearings splashed in. I wasn't going to be beaten, so I filled the sump with water until the oil rose high enough to get into the trays where the big end bearings could splash in. That was one of my first makeshift repairs other than bolting the tyres. It worked out quite well, with the oil floating on top of the water. We continued back home without any further damage.

Much later, Bill Petrick, who was involved in mica mining at the 'Spotted Tiger' mine in Harts Range came to stay with us when he was courting my sister Elsa. He told me the black substance in the rock was good manganese ore. So that was the beginning of my prospecting days!

Dad Returns and Mother Leaves for Treatment

Dad returned home, still weak from polio and only just getting around with two walking sticks and found Mother was very unwell. Although I was only about 12 years old at the time, Sergeant Stott of Alice Springs issued me with a special driver's licence (Number 6 in Alice Springs) so that I could legally drive Mum and Dad to Alice Springs. The funny part is that our Dodge was the first privately-owned motor vehicle in Central Australia and the only other vehicle was a Model-T Ford owned by Sergeant Stott himself, so there weren't too many problems with congested traffic! When he was driving his Ford, Sergeant Stott talked to it as if he was still driving a team of horses. However, he was a stickler for the 'letter of the law' and insisted that if I was driving I had to have a driver's licence, but my age didn't matter!

Not long after Dad came home, Mum's health reached a breaking point and she became too ill to carry on at home. She had developed pleurisy and tuberculosis was also suspected, which she may have acquired from nursing some aborigines who died from TB around that time. The sisters at the A.I.M. Hospital looked after Mother for a while, but there were no doctors in Central Australia and she ended up needing treatment in the Royal Adelaide Hospital around late 1927 or early 1928.

Our friend, Pastor Kramer, the Baptist missionary, looked after our home and family while Dad took Mother, Elsa and Martyn down to Oodnadatta and put them on a train to Adelaide. Dad then returned to Deep Well. One can imagine the physical pain and distress that trip must have cost both our parents. Elsa took up a position in Adelaide as a housekeeper to an elderly woman so that she could be near and visit Mother regularly. When Mother was finally discharged from hospital she spent some time recuperating and visiting her family in the Barossa Valley before returning home to Deep Well when she had sufficiently recovered. Elsa and Martyn had

Father, Kurt, Trudy, Elsa (back) and Martyn, Randle, Mona and Mother holding Myrtle at Deep Well, August, 1927. Photograph taken by the RESO Party (Photo courtesy State Library of the Northern Territory, Northern Australia Collection)

gone to the West Coast for a time and also returned to Deep Well with Mother at the end of 1928. Elsa helped our family pack up for the move to Alice Springs.

Thinking back over my childhood days, I didn't at the time realize how tough things were during the 1920's. First we had experienced the floods of 1921, followed by widespread bushfires the next year as a result of all the extra growth. This was followed by the 1922-1929 drought, with frequent electrical sand storms which caused the sand dunes to encroach everywhere, covering the stockyard and fences in front of the house. Barrow loads of red desert sand were carted out of the house after each dust storm. Large clouds built up every couple of weeks and became electrical dust storms and if it rained we got 10 points of mud! Summer temperatures were around 100 to 110 degrees Fahrenheit (40 to 45 degrees Celsius) in contrast to the extreme frost and cold caused by the drought in the winter, with half an inch of ice on the cattle trough. All these things were accepted as a normal part of life to us children, as they were to most other early settlers' children in 'The Centre'.

The pioneer pastoralists of yesteryear and their loyal women really worked hard for the pittance received, to rear and educate a family and build up a home. Often they lost out to flood, fire, drought and sickness. But if there was anything left they would bounce back and start all over again with all the optimism and courage in the world. However, drought and sickness had taken a heavy toll on our family and on Deep Well Station. It virtually bankrupted Dad, so we had to leave Deep Well and move to Alice Springs.

Myrtle, Kurt, Randle, Mona, Trudy, Mother and Father at Deep Well Station shortly before our move to Alice Springs, 1928

PART II
PRE-WAR DAYS (1928–1939)

Chapter 7
Our move to 'The Alice'

Deep Well Station was abandoned by our family early in 1929 following the devastating effects of the long drought years and the illness of both my parents. Sixty years later (1990) a few posts of the stock yard stand rotting in the centre of the sandhills and the crumbling walls of the stone house Dad built are still there, but the timber and iron are long gone.

Kurt standing near the ruins of Deep Well and the old homestead, 1989

The railway from Oodnadatta to Alice Springs was still being constructed past Deep Well during our move to the Alice. Ultimately its construction had quite an effect in changing the way of life for people in 'The Centre'. It meant the beginning of the end for the camel trains and also the end of the mail trucks and general provisions which wound their journey up the main south road – the famous 'Oodnadatta Track'.

The first train going past Deep Well during line construction in 1929

Dad had been forced to declare bankruptcy and the Bailiffs took over the station. They had sent a man out to take stock of the station when it was realized that Dad was unable to carry on out there. We owed the Bailiffs about £2,000 but they allowed Dad to keep his 1924 one-ton Dodge truck. The 1922 Dodge utility was possessed by the Bailiffs, which I later bought back from them for £10 so that the family would have a second vehicle to help earn our living. Both vehicles proved invaluable for the various contracts Dad and I later obtained.

Prior to our move, Dad had built a large shed at the back of a block he had leased at 51 Todd Street, Alice Springs. Mother and Dad gathered up our few possessions and took Mona 5, Randle 3, Myrtle 2 and Elsa's son, Martyn 4, to the block where we all lived for about four years while the house was being built. The shed was divided with canvas partitions into a kitchen and living room area with a bathroom at one end and the boys' room was at the other end. Mum and Dad and the girls had a tent each.

Elsa, Trudy and I stayed on at Deep Well until the 250 horses and cattle remaining, out of a former muster of about 800 before the drought, had been sold by the Bailiffs (Bennett and Fisher) and delivered to their new owner. In comparison, the Hayes family who had 12,000 head of cattle before the drought, finished up with only 800 or 900 head of cattle. Their main problem had been a lack of permanent water for the cattle. They only had three or four man-made waters and relied on rainfall and so-called 'permanent' waters which had dried up.

Until our debt was paid off, all contracts and earnings undertaken by Dad had to be registered in my name. This was to protect our earnings under the rules which applied to Dad's undischarged bankruptcy. This made things even harder for our family throughout the Depression years in the 1930's until all our creditors had been paid.

Our first Christmas in Alice Springs in 1928 was celebrated together in our new 'home'. We all attended Mr and Mrs Norman Jones' renowned Christmas Eve Party, which had been held annually since she started the tradition in about 1920. It was held in the back yard of their home in Todd Street, on a clay tennis court, beneath some shady gum trees where Ansett now stands. Tarpaulins were pegged down over the ground and dusted down with boracic, which made a smooth, slippery finish suitable for dancing to live music.

Mrs Windle played the piano, Jimmy Lackman played his steel guitar , Musty Milnes the violin if he hadn't got into the 'Christmas spirit' too early, while Bill Colback played the saw or his saxophone or accordian; and sometimes phonograph music was played. The sides of the yard were covered with branches from gum trees and decorated with coloured streamers and paper lanterns lit by candles.

Father Christmas arrived dressed in the traditional red and white garb, but sometimes arrived by quite non-traditional means, such as camel! Gifts were presented to all the children and any bush children who couldn't attend the party received their gifts on the following mail run. That year Bill Petrick was Father Christmas and he arrived on camel. He had been courting Elsa since they met in 1927 with Bill riding in from the 'Spotted Tiger' mica mine in Harts Ranges to see her. They married in November, 1929.

A sports meeting, lasting all day, was held on Boxing Day. All the events were held in Todd Street which was a wide dirt road in those days. There were dozens of races and competitions for both children and adults, including high jumps, long jumps, hop, skip and jump, bag races, sprints, egg and spoon races and anything else they could think of. The entry fee was two shillings per person. Most entry fees for the children were donated by bystanders and the winners for each race got the lot. The person who came in last would collect a booby prize, which was always strongly contested when the name was announced.

Getting Settled in Alice Springs

Before building a house, it was essential to sink a well and install a windmill and engine pump to establish a water supply for the vegetable garden and lucerne patch for our milking cow and chooks. Except for a few government-built houses, the homes in Alice Springs at that time were mostly timber and iron shanties. The government buildings were constructed using concrete blocks. Although there was an abundant sand supply, every bag of cement had to be transported in by camel train or by truck from Oodnadatta before the railway line was completed. The cost of cartage was as much as the materials and labour, which made it too expensive for the average person. Many people had to settle for bush timber and iron dwellings, some with thatched roofs.

By 1930 Dad was able to start building our house. Dad and I, with a couple of aboriginal helpers, quarried the stone and lime, transporting many a heavy load back home on the truck. We burnt the lime in a kiln we made near Heavitree Gap. Ten aboriginal prisoners, guarded by a warden, were available for hire at thirty shillings a day for the ten and helped to dig the cellar. Our house was built of stone and lime mortar, with wide shady verandahs.

While Dad worked on the house starting at the back, Mother planted a garden in the front and side of the house. She planted a beautiful display of flowers including roses and pansies on the northern side. Date palms, grape vines and fruit trees were also planted and later provided abundant crops for the family. Later, when the vines were producing fruit, she sold some the grapes to bring in a bit of extra money. A pound of grapes sold for the price of a loaf of bread. When the house was completed the two front rooms were let to boarders to supplement our meagre income.

Our house was sold to Connair for Staff quarters when Dad died in 1951. It was demolished in the 1980's and the National Australia Bank building was erected on the site.

A view of the railway station and new Post Office in Alice Springs, around 1934

Our home in Todd Street, Alice Springs, built around 1930

Some more Schooling for me

The day after my fourteenth birthday, in 1929, I started attending school in Alice Springs. At that time the school was a tin shed one-roomed building for the fifteen to twenty children of all grades from first to about seventh grade who attended. Our teacher was Miss Burton, a small bird-like spinster. I started in grade three, as I had previously completed grade two by correspondence lessons.

It was often extremely hot in the tin shed but we weren't really used to anything else, and I didn't mind it very much. My first 'heart-throb' was a girl named Doreen Martin, daughter of one of the earth-moving contractors who came up north with the railway construction crew, using Clydesdale horses and scoops. I was pretty sweet on Doreen! She was short and plump, but she had the sweetest little 'cupie-doll' face, and I fell in love with her. I always wanted to sit by her and talk to her, even if we were teased. She didn't like the teasing much and would buzz off when it started, but I never took that to heart.

One day Len McDonald, a couple of other lads and I were playing cricket in the yard near the girls' toilet. The girls' and boys' toilets were just five posts with some hessian tacked up around them. I made a swipe at a ball and missed. The bat slipped out of my sweaty hands, flew up into the air and sailed over into the toilet, landing right on Doreen's shoulder. Another girl, Queenie Sullivan, who could swear and curse in really rough and tumble style, was in there too and came racing out to accuse me of doing it on purpose.

Teasingly, Len agreed that I'd thrown it in there on purpose, which started my one and only 'box-up' fight. We were fair dinkum, fighting away and starting to bump up against the tin chimney of the school. Miss Burton raced out and started dancing in between us like a rooster between two hens! We saw the funny side of it and burst out laughing. Miss Burton was not amused and took us both inside and administered six whacks each with a ruler over our hands, which made us laugh even more. In the finish she was so furious and frustrated she sent us home!

By the end of that year I completed and passed grade three work, but the family needed me to help earn money because Dad's legs were still very weak as an aftermath of the polio. Although I was only fifteen years old, I was tall and strong, had plenty of energy and knew all about hard work. That was the end of my formal schooling.

A Close Shave for the 'Faith in Australia'

The aeroplane 'Faith in Australia', which was visiting Alice Springs in about 1929, had a close shave from coming to a fiery end at my hands. It was the sister ship to two other famous aircraft, the 'Southern Cross' and the 'Southern Cloud'.

I was about fourteen years old at the time and had just acquired my first pair of long white trousers and was feeling very proud of them, too! I was being paid ten shillings to look after the plane for the night, so I bundled up my swag and went out there to sleep inside it.

Sir Charles Kingsford Smith's "Southern Cross", sister craft of the "Faith in Australia"

All things mechanical fascinated me and I had been hankering to know more about aeroplanes since seeing Francis Birtles fly over our home at Deep Well several years earlier. I was very curious about the altometer and kept on lighting matches to get a better look at it up near the ceiling in the passenger cabin. I was holding a match up near the altometer when, to my horror, the doped fabric above started to sizzle and burn! Panic! Dread! Quickly, I rubbed it out with my fingers. I certainly didn't light any more matches that night.

Next morning I helped start the plane by winding the inertia starters. As I was doing this, a mixture of black oil and petrol trickled down from the engine and dripped all over my lovely new whites. The aeroplane had its revenge!

Chapter 8
First Work and Early Romance

Odd Jobs and the Night Cart

From the age of fifteen onwards it was all work and I became another bread-winner in the family. Dad let Trudy and me keep five shillings per week out of any money we earned, the rest of which went into a consolidated account for the benefit of the whole family. From that amount we had to supply our own clothes, shoes and anything else we wanted, although five shillings in those days had good purchasing power. It was more like an allowance really, which we older members of the family received. Everyone, even the younger children, still had their jobs to do around the home. They had to help me early in the morning with milking the cow, feeding the chooks and cutting lucerne for the livestock.

I started in transport right at 'the bottom', you might say! I ran the first night cart and garbage collection service in Alice Springs, using the Dodge 4 truck. There were only 23 deliveries in 1930, which shows how small the town was then. I continued the sanitary and garbage service contract for about four years for which I was paid £24 a month. Dad and Bill Petrick, my brother-in-law, stood as sureties for me since I was under-age to hold a government contract. Our family earned enough money to meet our needs, but nothing to spare.

Alice Springs in the early 1920s with the A.I.M. Hostel (Adelaide House) centre-left and the old school I attended centre-right, standing alone.

Although Dad couldn't do much physical work he still ruled the roost with his stockwhip. If anyone stepped out of line he'd better look out. The last hiding I received was one Sunday afternoon when I was fifteen years old. I'd gone over to McDonald's place, the station master's home, and stayed there chatting with my friend Len. Mr McDonald was practicing on the piano with Stan Cawood playing a violin, which I was enjoying very much. I came home just before dusk, at about eight o'clock, and there was Dad waiting for me at the gate with the stockwhip. That was the last time he used it though, because I was getting too big and I overheard Mother warning him that he might 'lose' me if he kept on treating me as a child.

Besides helping with upkeep on the place Trudy started a photography business taking photographs, developing film, printing, enlarging and tinting. She set up a dark room for that down in the cellar. Dad and I carried on with general work, which from 1930 to 1933 included odd jobs, building houses, putting down wells or bores and carting firewood for the bakery, hotels and boarding houses. I also did a bit of mechanical work, repairing old trucks, windmills and pumps.

I rigged up a charging plant with an old Lister engine and generator, supplying 32-volt power for lighting three houses – ours and two neighbouring houses. I also charged batteries for the old-type radios, charging two shillings each. It all helped to earn a bit more money.

One day I had just finished repairing old 'Mum' Meyers funny old 1918-model Chev truck. She asked, "Will you take it out for a run and bring me back a load of wood?" I agreed and another young lad, a friend of mine, came with me. We drove out over the Todd River about four miles out and started loading up some wood when a thunderstorm struck. The rain virtually flooded us out. The old truck refused to go as it was all open and wet (and so were we!) so we walked back into town and told Mrs Meyers what had happened.

That afternoon, after the rain cleared, Mrs Meyers, who also owned a flash new Graham-Paige car (which she couldn't drive herself) said, "Well, you drive me out and I'll come too for the run. We can pick up the truck and bring it back, and your friend can drive me home."

We went out to collect the truck but when we came back the Todd River was in flood. We had to leave the two vehicles on the eastern side of the river. I suggested, "We'll have to wade across!" When we went in the water turned out to be about 5 feet deep! Luckily Mrs Meyers stayed calm. None of us could swim, but we were quite used to playing in the river when it was in flood, floating along with the stream. We took her short, dumpy form by one arm on each side and practically floated her across. We just touched the ground every now and then and gradually pushed our way across the river. About half a mile further downstream we finally reached the other side without even a thought about the dangers of it. Thinking back now I can realize the danger if Mum Meyers had panicked – it probably would have been the end of the lot of us.

The Radio and the Races

Len McDonald Senior, the first Railway Station Master in Alice Springs, was also a radio ham. He and I used to tinker around building radios in our spare time, making up many of the parts ourselves. One set was quite powerful and could pick up Adelaide and Melbourne during the day, which at that time, was almost unheard of.

The Todd River flooding Alice Springs

At that time, in the days before trunk lines, the local SP bookies received the results of the southern horse races by telegram, which meant that the results would arrive about an hour after the race was run.

One day a chap known as 'Snowy' heard about our radio and decided to check it out. When we told him the set had been delivered to Stan Fleming, one of the bookies (for whom it had been built), Snowy showed great disappointment and asked how much it had cost. We told him £55.

"Christ!" he exclaimed. "I would have paid £100 for it!" and stamped off.

I asked Len, "Where would he get £100 from?" because Snowy was an odd-job man and it was during the Depression, when things were pretty tough for everybody.

Snowy, however, was cunning. The following Saturday he went into Stan Fleming's betting shop and put five shillings on a horse. The horse lost but he listened to the race and the winner came in at 20 to 1. Snowy walked out of Fleming's shop and raced down to the other bookie, where he put £10 on the long shot.

Baker said to him, "You're mad! You'll do your dough!"

Snowy replied, knowing the horse had already won, "Well, if I do, I can square you up from that windmill job I've got next week."

Snowy went straight back to Fleming's shop to listen to the next race and put on another 'dummy' five bob bet on a 10 to 1 horse, which lost. The winner, Black Knight, started at 6 to 1. Snowy raced back again to Baker's shop and asked him to put the lot on Black Knight. Baker said he'd deduct what Snowy owed on the book and put the rest on, which was £140. Snowy did one more similar stint on the last race.

That evening, Stan Fleming and Baker were having a few drinks at the pub, discussing the afternoon's racing. Stan raved on about his new radio and how he got every race straight from the course and cleaned up on most of the races. Baker said, "I would have been okay too, if it hadn't been for Snowy. He cleaned me out for nearly £2,000 and I'd like to borrow £1,000 from you, if you could manage it."

"But Snowy was around at my shop most of the afternoon, having five bob bets!" said Stan.

Then the penny dropped. Baker was a hot-tempered bloke who ran a barber shop along with his betting shop. By this time he'd had a few drinks under his belt to drown his sorrows, and was really mad when he realized how he'd been cheated.

"I'll kill that bastard, and you, and that f------ McDonald and Johannsen too!" he yelled.

Stan Fleming was a fairly big, flabby-looking man with a florid complexion and a placid, friendly nature. After this outburst by Baker, Stan's face started to turn purple and he tried to escape from the corner of the bar where he was trapped.

At this point, Dick Turner came in, looked them up and down and remarked, "You two look about as happy as a pair of 'horny' Jersey bulls that are about to be castrated with a hundred heifers waiting in the next paddock!" A great roar of laughter went up in the bar. Dick was a tall man, six foot three inches high, very strong and had a head of prematurely grey hair and friendly blue eyes twinkling behind the thick glasses which he wore. Dick, Alf, Tom and Jim Turner, four brothers, first came into 'The Centre' in the early 1900's and were well-liked around there. Alf, Tom and Jim were cattle men, while Dick was mostly interested in mining. He was a heavy gambler, but very honest.

Dick listened to what had happened and told Baker to "Cool it" and go and ask Snowy for the money back. He offered, "I'll come with you, but keep your shirt on! Now, meet me here in the morning and we'll get the money back off him. You're too f------ drunk now to do anything." Dick escorted Baker to his room with a bottle of beer for a 'heart starter' in the morning, laid him on the bed, took off Baker's shoes and went back to the bar.

At ten o'clock that night, as the bar closed and interval lights came on at the open-air picture theatre across the street, all hell broke loose. The exit doors crashed open and people came running out with women and children screaming, and men yelling – then dead quiet.

Baker had got up from his bed, gone to his barber shop, selected a 'cut-throat' razor, tucked it in his hand and gone to the Theatre. He walked up the centre aisle until he spotted Snowy, then sidled along the seats behind him, grabbed his head, pulling it back with one hand and slashed at his throat with the other. Snowy was wearing a trench coat with the collar turned up, which probably saved his life, but it still took 120 stitches to sew him up. Paddy Riley, the local doctor, was attending the pictures at the time and stopped some of the bleeding which was spurting everywhere. Luckily, the A.I.M. Hostel, which was the hospital, was only 200 metres away.

Snowy never laid charges against Baker but I think he was sentenced to four years in prison for malicious assault. Snowy never looked quite the same again after that. He went up to Darwin and later started trading between Timor and Darwin using a barge from Ex-Army Disposals.

First Motorbike Race at Alice Springs

Jimmy Lackman, Billy Fox and 'Bronco' Smith were some of the stewards at the 'off-day' race meeting which was a day for unofficial races and events held at the race course the Sunday after the official events. They asked me if I could organize a motorbike race for the last event of the day. It would be between Hughie Clough on a "Sunbeam", Jack Hughes with his "Indian Chief" and side-car and me on an old 1926 "Norton". This would have been about 1932 when I was seventeen.

The race started and I was leading during the first lap, amid cheering from my backers, when suddenly my bike conked out. I fiddled around, got it going again and passed the other two riders again. My bike conked out again! When I'd been working on it the night before I'd loosened the tappets for adjustment and hadn't tightened them again. Anyhow, I managed to finish second. I actually pushed the bike over the finish line, receiving loud cheers from the crowd. It was all good fun and was the first motor bike race ever held in The Alice.

A Spot of Early Romance

When I was about sixteen or seventeen I decided my hair was getting a bit tangly. I couldn't wash it properly when I was out bush carting wood, digging wells and so forth and I'd seen another person with his hair shaved right off so I thought I'd do the same. I took the hair clippers and ran it right through the middle, then went to show my sister Trudy, saying, "Look what I've done!" Trudy mowed the rest of it off for me. Mum and Dad told me off for doing it, but that was okay by me.

A couple of days later the fortnightly train was due in and I was assigned the job of picking up the new governess and the Albrecht family and driving them to our home. Pastor Friedrich Albrecht was the missionary out at the Hermannsburg Mission by then. A young girl, about the same age as I, stepped off the train along with the other visitors. There I was, with no hair, meeting her for the first time! Her name was Ruth Pech. She looked at me and had a bit of a giggle.

Before I knew it I was smitten. She turned out to be the first real heart-throb after Doreen that I had. In fact, it was young 'puppy love' for both of us. Any opportunity I had, I'd go out to Hermannsburg and stay there a few days. If I took a load of stores out I'd stay a couple of days, and occasionally she would come into town.

One thing worried me though. She was a very serious and dedicated Lutheran. She told me her ambition was to be a missionary and go up to New Guinea. Questions were asked as to whether I'd like to be a missionary and go up there too. Of course, that was right out of my line. I didn't

The first official train arriving in Alice Springs on August 4, 1929 with our Dodge 4 on the left

mind missionaries but as far as *being* one, no way! I broke off our friendship and said I wouldn't see her any more because I could never be a missionary. It was easy to *say* at the time.

I returned home and went down to Deep Well to help Dad who was in charge of sinking a well for the Railways. We camped there for about a week and my nights and days were absolutely painful. All I could think about was my love over at Hermannsburg. I'd never realized before how painful love could be. I just wanted to run back to Ruth and say, "I'll do anything!"

Not long afterwards, another young lady and several others arrived in the Alice wanting to be driven out to Hermannsburg. First they wanted to be taken around sight-seeing and I was given the job of taking them as tourists out to Standley Chasm, Simpson's Gap and other places. Marcy Homburg and her friend, Mrs Menz, stayed at our place. Marcy was quite a sophisticated young lady, three years older than I. When Mum and Dad noticed that we were showing interest in each other they 'tabooed' the idea saying, "But dear, she is so much older than you! You couldn't possibly show any interest in her." At any rate I took the visitors out to Hermannsburg and to Palm Valley and found by then that the terribly strong attraction I'd had towards Ruth had softened considerably with the arrival of another lovely young woman on the scene. The flame had gone out and there was only a bit of smoke there!

The visitors returned to Adelaide and about six months later Trudy and I went down for Christmas and stayed with them at Glenelg, but that was the end of my second romance. Her father was a Q.C. lawyer and their way of life was vastly different to ours. They loved visiting our way of life and we loved visiting their way, but as far as living the same way and changing lifestyles it couldn't have worked. Marcy was a great tennis player and into sport, dancing, art and socialising. About 40 years later I bumped into Marcy again. However, my second heart-throb had not been as painful as the first.

Around 1934 Mum had a young girl, aged about twenty, working for her doing general housework and helping. I was nineteen then. Nobody else was at home one day and we started 'mucking around'. She got a bit frisky and suggested she could teach me some things. At that time I didn't know too much about sex and was too frightened to try anything. She suggested, "Well, Chinaman Sing Fong sells 'French letters' so you could go and buy one and then I'll show you." But no, I couldn't. I think she eventually bought one herself and next time we were alone we tried it out. It was all over before we even started as far as I was concerned because it was all too quick! After that she gave me the 'sack'. She reckoned I was no good at all – I didn't know anything. So that was the beginnings of my sex life.

Chapter 9
First Mails and Mining

Mail Contracting

Dad and I always kept our eyes open for additional work, since the sanitary and garbage collection didn't take up very much of our time and we needed to increase our income. The population of Alice Springs, or Stuart township as it was then known until 1933, was around 200 at the time.

In 1932 we took over the mail contract east of Alice Springs from Sam Irvine. His mail run had only gone to Arltunga, but ours was extended to take in Winnecke, Ambalindum, Claraville, Arltunga, Mount Riddock, Alcoota, Waite River, Macdonald Downs, Mount Swan and Delny.

Arltunga and Winnecke were small gold-mining settlements which had been reasonably prosperous in the late 1800's up to 1914, when the population dropped to about 20 miners still scratching out a bit of gold. During World War I the use of explosives was restricted which limited the mining operations, as had the call to service of the young able-bodied men.

The vehicle we used was the 1922 Dodge 4 utility from Deep Well, which served us well around Alice Springs and on the Arltunga mail run. It had many modifications by then and was ideal for negotiating the rocky, rough stretches on primitive tracks.

My First Encounter with Alec Kerr

Alec Kerr, the owner of Delny Station, was an eccentric, wild Scotsman who lived with his sheep, literally. He camped in the sheep yard and didn't like his 'wee lamby-lambs', as he called them, being disturbed by motor cars coming onto his land.

On my first mail trip out there, upon my arrival at Alec's place, I was confronted by this barefoot man with a scruffy beard and shining, piercing eyes, pointing an ex-World War I .303 rifle at me. He yelled in his Scottish accent, "YE! Ye arrre on my land and I will shoot if ye come one inch furrrrther!"

I had been warned that he might be dangerous, but didn't quite expect this apparition dressed in ragged trousers and shirt, shiny with filth, and a hat fashioned out of a wire rim with hessian bagging sewn onto it.

After what seemed an eternity, I managed to croak out, "I've brought your mail! I am the mailman – the Royal Mail." And I handed him a mail bag with his name on the brass tag and 'P.M.G.' stamped on the bag. He looked at it, took it, and slowly lowered his rifle.

I was also carrying some stores for him from Wallis Fogarty's, the local store and general agent in the Alice. His order included a bag of flour, 150 lbs (pounds), 70 lbs bag of sugar, 10 lbs of tea, 6 tins each of baking powder, golden syrup, jam and a bag of salt. In a crackling, sharp voice he thanked me and told me I should now leave, slowly, being sure to follow him.

He said, "I'll show ye wherrre to go and in future use the same trrrrack. I have put shear blades in the other trrrrack to cut my enemies' tyrrrres!" and, giving a hideous cackle, he sent me on my way.

I was never so glad to leave a place! On later trips though, he became quite chatty, always telling me about his enemies. They ranked, apparently in order of how much he detested them, dingoes, Sam Weller from the mica mines, some neighbours, wedgetail eagles, aborigines and travellers.

Some Missing Money

On one of the eastern mail trips in 1934 an Italian who had been in town to sell mica came along with me as a passenger. Five or six of his partners had placed him in charge of selling the mica and he was bringing the money back to them. The Italians out at the mica mines seldom used banking accounts in those days as they didn't trust the banks. Apparently in Italy some banks had collapsed and people had lost their money. These fellows hid their money and all their transactions were on a cash basis.

At Alcoota he took his little portmanteau out and rested it on the back of the utility to get something out of it. When we left he had forgotten that it was still on the back of the ute. We had driven on for quite a few miles, almost to the next station, when he wanted some tobacco and asked me to pull up. He looked for his portmanteau, but it wasn't in the truck. His face turned a greyish-white.

I asked, "What's wrong? Are you sick?"

"Yes," he replied, "I'm sick all right. I have lost my portmanteau!"

"Is there anything valuable in it?" I queried.

"Yes!" he answered. "Over £4,000 for paying my partners, and if I don't turn up with that money I will have to go bush because they will shoot me!"

We turned back straight away and retraced our journey keeping a sharp look out for the missing portmanteau. Three or four miles from Alcoota, there it was, lying on the side of the road. He was a very happy man!!

The Italians out at Harts Range were good customers of mine, for both the mail and extra loading I carried for them when they needed barrels of wine, general stores and fuel. I also carted their mica back to town on the return trip if they had any ready. The rate for loading in those days was sixpence a ton per mile which was just a little above costs, but every bit of extra money helped. The mail contract alone only just covered petrol and running costs, so any extra loading we could obtain gave us a margin of profit. Occasionally I had a paying passenger, but most of them were non-paying. However, it was a really friendly run around to all the eastern stations where I knew everyone personally. There was a real family feeling about it.

The Bales of Wool

On one trip I had some bales of wool to pick up from Bill Petrick, my brother-in-law, at Mount Swan. I was rather nervous about it because on the way back I had to pass Blackfellows Bones Hills, across from Mount Riddock through to Claraville and Arltunga, which I knew was a really rough 'goat track' of a road. At one place I was extremely nervous about continuing along a steep slope where the track ran along the side of the hill for about half a mile. My top-heavy load was hanging over so much that the high-side back wheel couldn't grip and just kept spinning around. By clearing a bit of a track down the hill and along one of the ti-tree gullies, I managed to dodge the worst part of the slope. It took me half a day to clear and make the new track along the bottom and then go up a steep jump-up. After my road-making effort that became the normal track and was subsequently used by everyone travelling along there.

When I pulled up at Claraville old Larry Rosenbaum couldn't believe that I had carried such a heavy load through that stretch. He was a funny old character and many people were frightened of him as he had the reputation of having a helluva bark, but his bite wasn't bad at all. He relied on the mail service to bring his gear out and always treated me very well.

Mrs Cavanagh of Ambalindum (which means "little child has walked away and become lost") was another who had the reputation of having a strong bark and her bite could also be pretty sharp! Most people preferred to deal with her husband Fred, who was a real gentleman. The house was completely out of bounds as far as any workers or strangers were concerned. However, at the workers' hut Fred would hospitably offer, "Would you like a cuppa? We're just boiling the billy."

Kurt in Mulga Express Mark I, just returned from a mail run to Mt Swan with a load of wool bales, 1936

A Black Spot on the Record

From Ambalindum we went across to Arltunga where there was a government battery gold crushing plant. A less pleasant character along the run was the policeman there, George Murray, who was in charge of the eastern district. He was rather crude and very tough, especially on the aborigines. The Government had brought in a policy of rationing for the aboriginal population for the benefit of the old, infirm and the children. All police stations were allocated ration points according to the numbers in the district, so that the aborigines could come in and collect weekly rations either from Alice Springs or their nearest centre. The rations were intended to be free but Murray was alleged to have been an "aboriginal hater", and I was told by Walter Smith, who had lived in the area for many years, that out there at the goldfields Murray refused to issue rations unless they brought in a bit of alluvial gold. This meant he was actually selling the rations to them.

'Nugget' Morton, who owned Ammaroo Station, and Murray were also allegedly involved in the 'Sandover Massacre' where 100 or more aborigines were either shot or poisoned after it was alleged the aborigines had speared some cattle. Apparently strychnine was put in the soakage of the Sandover River. Murray was also involved in the 'Brookes Soak Massacre', so-named because an old prospector named Brookes who had camped at the soak was speared while sleeping in his swag. In this instance it was alleged Brookes had taken one of the aboriginal women without the consent of the tribal elders. The massacre was an act of revenge. The exact number of how many aborigines died was never officially given, but those who lived out in that area say that one can still go into certain areas and find hundreds of bones scattered around in the scrub.

An Independent Bloke

On the eastern mail run we travelled from Arltunga back to the Garden Station of Jim Turner and from there up to the Winnecke gold fields. An old timer at Winnecke was Maurie Thomas who was a very independent pensioner. He fossicked for a bit of gold and would often push his wheelbarrow 48 miles (80 km) over some very rough, hilly country into Alice Springs. If I met him along the way he usually refused a lift. Eventually he bought a push bike and once a month he rode into Alice Springs to collect his pension and a few stores.

He had a fabulous garden established near Winnecke Well. He pulled the water from the well by windlass from about 80 feet (24 metres) down and poured it into a small tank beside the well. He then put a yoke on his shoulders and carried two four-gallon buckets about 150 metres to his hut to provide for his personal needs and to water his garden.

One day I met Maurie on his way into Alice Springs, laboriously pushing his bike along. I asked, "Why don't you put it on board?"

He said, "I will, on one condition. I want to pay you for the ride in. I'm not feeling very well and I will accept the lift but I want to pay you."

I answered, "Well, seeing how you are always mending the road by filling in wash-outs and clearing the rocks away while you are down the track prospecting or doing a bit of hunting, you are actually paying your way by improving the road and it is very much appreciated."

He accepted that and from then on he really went to work and made miles of new tracks in the next twelve months, which improved the road no end.

A Trickster being Tricked

Len Garland was another character out at Winnecke. He originally came into the Territory as a salesman for forestry bonds, which were supposedly 'shonky'. He may have been caught up with by the authorities and had possibly gone out to Winnecke for several years to hide away. He drove a big Chrysler with buffalo horns mounted on the front bumper.

He discovered quite a good alluvial gold show, working with 'German Joe' and Norman Crowther, supposedly in equal shares. The others did the digging and carting while Garland extracted the gold with the shaker at Winnecke Well. When the other two were digging they unexpectedly found a very rich patch and obtained quite a lot of gold in their sampling dishes. They kept some of the gold out of the sampling dishes and just hung onto it for a while, but they continued to send a truck load of the same dirt down to Len every day.

A couple of weeks passed and the results at Garland's end were always the same – about half an ounce a day, whereas at the digging end from their sampling they knew they should be getting about two ounces a day. They suspected that Garland must be salting it away and not letting on that they had struck it rich. At that stage they decided that if he was going to cheat on them they would cheat him. They started to pan a lot more of the rich ore and finished up extracting about 80 ounces of alluvial gold before the patch cut out.

They decided then to "give the mining away" and told Garland they'd "had enough of this digging". Each wanted to take out his share so Garland divided up some of the gold and the others left. They hitched a ride back to Alice Springs on the mail truck with us and the first thing they did when they got into town was to celebrate with a few beers. Of course, Old German Joe drank a few too many, became very talkative and their secret dribbled out. He told some people in the pub that they were going up to Halls Creek the next day to sell the gold on pretext that they had mined it from up there.

They left and the following day Garland came in looking for them, having figured out that they must have had a lot of extra gold. He was told what German Joe had said in the bar so he sent the police after the two men. They were apprehended at 'The Granites' and were brought back to face charges. However, on a technical matter of law they were only fined a small amount because a minority cannot impose itself upon a majority. So Garland lost out.

My First Mining

As the mail run was only a monthly service there were times when I was available to take tourist parties or expeditions out, which is described later. Also, during the mail runs out east I became very interested in the gold mining at Winnecke. In between mail runs I started gold-digging in the 'Black Eagle' shafts which Webb Brothers had previously worked before World War I. Webbs became the owners of Mount Riddock Cattle Station in the Harts Ranges. They told me there was still some rich ore left in the 'Black Eagle' mine.

Eventually I put in a small crushing plant and four or five of us (myself, two aboriginal men and two half-caste chaps – Jack Hughes and Wauchope Tilmouth) worked it. The crushing plant produced one ounce of gold a week, at that time worth £9, which was a good supplementary income. My workers were paid ten shillings a week, plus board and keep, and the rest of the money was put into the bank in the family fund.

Dogged by Dingoes

Early one Sunday morning in about 1932 or 1933, during the time I was working at the crushing plant at the Black Eagle, I drove out towards the Nine Mile mine before Patsy Ciccone had set up there. I went as far as Nine Mile Creek where I began prospecting up in the hills beyond the creek. About two o'clock in the afternoon while I was sitting on a rock having a bit of a spell, I noticed a dingo nearby. I walked on further and a little while later I saw another dingo on the opposite side of the pad I was following. I reached the place I intended to prospect and had a look around. I found a few colours of gold but nothing worth working or worrying about.

I set off to return to the truck and as I walked along the pad, out of the corner of my eye I saw a dingo again, about 50 metres to one side. I stopped and the dingo also stopped. I sat down for another spell and noticed another dingo on the opposite side. I casually looked around and saw yet another dingo behind me, making three of them. I wasn't too worried about it yet and continued walking. After I'd gone three or four miles I sat down again to rest, since it was fairly hot and I was getting tired. I looked around again and this time there were two dingoes – one behind me and another on the side. Just seeing them there constantly and realizing that they were following me suddenly made the hair crinkle up on the back of my neck. By the time I returned to the truck they were only about 20 or 30 metres behind me. It was pretty 'droughty' at the time and they were probably following me, hoping I would perish. They must have had it in mind there was a feed for them if I conked out away out bush amongst the hills, or they may possibly have attacked. It was the first time I had encountered that kind of behaviour from dingoes away out in the bush.

A similar incident happened a few years later when I was riding my motor bike from the hills in the south up to Patsy Ciccone's Nine Mile crushing plant on the opposite side of the range. I'd seen a bit of a saddle in the range and thought I might possibly be able to take a short cut through there, but I couldn't get the bike down through there, which was where I ran out of fuel. It was a very narrow gap with quite a steep drop. I had to walk around the point of the range to reach Patsy's camp. He only let me have a small amount of petrol and oil so I set off back to the hills again. I ran out of petrol again about 20 miles (32 km) from Alice Springs and made camp. It was winter time so I kept warm near my campfire during the night.

Early next morning I heard some dingoes howling in the distance. Shortly after hearing the howls a young kangaroo came sailing past followed less than a minute later by a pack of about four dingoes. They were spread out about 300 metres apart, following the kangaroo and wearing it down. It was obvious the kangaroo had just about had it. On a nearby wide open plain they closed in until the kangaroo stopped. They circled around, made a dash and got him.

Before I'd left home in the Alice I had told Dad approximately where I was going. When I hadn't returned Mum and Dad thought I must have broken down or had an accident so about nine o'clock the next morning Dad came along looking for me. I siphoned some fuel out of the Dodge and rode the motor bike back into town.

Another time on the mail run near Macdonald Downs I saw a dingo dragging a half-grown lamb weighing about 30 pounds. The dingo, quite a powerful dog, had no trouble carrying the lamb down into a creek. I presumed it must have a litter of pups nearby, otherwise it wouldn't have dragged the lamb so far, sometimes dragging and sometimes carrying its prey. Some people say the dingoes sling their prey over their shoulders like that if a load is too heavy to drag.

Dad's accident

Dad and Trudy also came out to Winnecke gold mine to help. Trudy stayed on for several months, doing the cooking for us and Dad helped with the crushing. Dad wasn't very enthusiastic about my gold mining at first, but later on he too was bitten by the 'gold bug' when he saw me smelting the gold on the forge. It's a great sight pouring your first slug of gold, even though it was only 8 ounces.

An accident happened one day in 1933 when Dad was walking around with goggles on to keep the ever-present flies out of his eyes. He was still a bit unsteady on his feet and unfortunately caught one foot in a belt of the machinery, which took it right around the pulley and badly smashed up his leg. Trudy and I put a rough splint on his leg and bundled Dad onto the back of the ute where we buried his leg in damp sand so that it would be held rigid. This was the best we could do to prevent it jiggling around, as we had to transport him over the very rough, rocky, bumpy road into town which was about 45 miles (73 km) away. It was a terrible trip because the old Dodge's clutch thrust had 'had it' and I couldn't use it.

It was another case of improvising and 'making-do'. By starting the motor with the crank handle, pushing the gear lever against the reverse gear, grating it a bit, bringing all the slack in the joints in the opposite direction, and then quickly flipping the gear lever into first with the motor just idling, it would take off. Then, by accelerating and changing gears and getting the motor revs just right so as not to crash the gears, it worked well enough through the many gear changes over the rough track until we arrived safely in Alice Springs nearly four hours later.

There was no hospital or x-ray facilities there at the time, only the Hostel run by the A.I.M. The doctor who was there set Dad's leg as best he could and put it in plaster. About six weeks later he took the plaster off to have a look and found the leg was about two inches shorter than the other due to the bone having splintered so badly. He had to break it again and pull it apart to re-set it. Dad had to endure this process once again. The doctor was doing the best he could without any x-rays, including putting it in traction. It was several years before Dad could walk on it properly. He used a walking stick, but that leg always remained about an inch and a half shorter than his good leg. It was another set-back following his efforts to recover from the effects of polio.

Through all this time Mother nursed him again at home which was another trying period for the family. I was away on mail runs and transport trips most of the time and she still had the three youngest children aged 9, 8 and 6 years old to cope with. Fortunately, Trudy 21, was still living at home then and helped around the house, besides operating her photographic business. After Dad's accident I had to abandon the gold mine and help look after the family in town. I took on the longer mail run to Tennant Creek and Birdum at this time and continued with general carting for the stations and mica miners in the Harts Ranges.

Mother and Dad with Mona, Kurt, Trudy, Randle and Myrtle 1936

A few years later, after his broken leg had healed and was stronger, Dad went out to the mine again with George Glass and Snow Kenna and sank a well where I had previously put down a bore. This increased the water supply and they treated the sands with cyanide to try and get the last of the gold out. They also mined a complex ore body at the Gander mine, about 3 miles (5 km) north of Winnecke. This ore body contained gold, copper, bismuth, antimony, lead, silver and zinc, which was quite valuable but it couldn't be treated in Australia and it was not an economical proposition to send it away to Germany for treatment.

First Mails and Mining

A Quick Trip to Tanami

Around 1934 a mining company which had some leases at Tanami had applied to the Mines Department for exemptions from working their leases. They were looking for somebody to go up and post the notices on the pegs. The Mines Department contacted Paul Johns who, in turn, asked me to run him up to Tanami in the Dodge 4. I didn't have much time available since I was due to take a mail run out in a couple of days. I told Johns we'd have to make it a really quick trip. (Paul Johns had arrived in Alice Springs around the middle of 1930 when Harold Bell Lasseter hired Johns and his camel team to assist in the famous search for gold near the Petermann Ranges in the Western Desert country. After an altercation out there with Lasseter, Johns had returned to Alice Springs while Lasseter is believed to have perished out in the desert after his camels had bolted, although some people still think he lived and left Australia).

Kurt atop an anthill in the Tanami Desert, 1934

Anyway, Johns and I left immediately, travelling on a bright moonlit night, going north via Pine Hill, Coniston, Brookes Soak and Mount Doreen, past 'The Granites' to Tanami. I felt rather nervous about going into that area because it wasn't long after the massacres of aborigines at Brookes Soak and the Sandover River areas. It was rumoured that the aborigines up north were quite hostile because many of their tribe had perished in the brutal poisoning of their water soaks.

However, we accomplished the job without incident and forty-eight hours later we were back in Alice Springs where Johns met the Special Magistrate, old Ben Walkington, who 'went crook' at him for not having yet left to go up and post the notices. He wouldn't believe Johns when he was told we'd already been up there and back in forty-eight hours. I had to go and convince Ben it was really true that I had driven Johns up there!

Passenger Services

As Dad's health and strength improved we established a passenger service which operated in tandem with the mail runs to Tennant Creek. Every Sunday at noon the mail coach, which was a Studebaker President 6, a nine-seater bus, departed from Alice Springs for Tennant Creek, 315 miles (500 km) north, calling in at Aileron, Ti Tree Well, Barrow Creek and Wauchope wolfram mining field, finally arriving at Tennant Creek at eleven o'clock on Tuesday morning if we were on schedule. We camped out two nights along the way.

On the monthly run to Birdum the passengers spent two more nights along the way, calling in at Banka Banka, Powell Creek, Newcastle Waters, Dunmurra, Daly Waters and arriving at Birdum by five o'clock in the afternoon on Thursday. The return journey commenced one week later to give Darwin people enough time to reply to mail. The south-bound schedule was similar to the northern one, which gives a fair idea of the state of the track in those days. Nowadays it would be a comfortable twelve-hour run.

The mail run to Birdum had originally been quite a good paying proposition when Sam Irvine had it because it was the only passenger connection between Darwin and Adelaide other than making the journey by ship. We'd no sooner got started when Guinea Airways commenced a weekly flight from Adelaide to Darwin. From then on my passenger traffic dropped to such an extent that I might only have one passenger per trip, whereas I'd previously averaged about eight. That run became a losing proposition even with the extra bit of general freight I carried. The mail contract on its own barely paid for running costs.

Also, during the Depression, there were quite a number of hoboes travelling around looking for work, staying in one place for a few days before the police moved them on. I'd average about eight or ten hoboes getting a lift up north or coming back down south but they were non-paying. Some of them were really decent chaps, but they were broke, just moving around trying to eke out a living. A few were 'professional' hoboes, but on the whole they were quite a decent lot. I ended up feeding about half of them but I'd limit rations to bread, jam and tea plus flour for making dampers. I always carried plenty of those staples in case of breakdowns or bogs on my trips up north.

Chapter 10
Trips for Tourists and Terrible Tracks

Besides the eastern mail runs Dad and I took on extra work driving for tourists or taking scientific expeditions out. Bert Bond (Bond's Tours) came up to Alice Springs around 1931 to do a tourist run. He started off on his first trip with two big, heavy old 7-passenger Studebakers. He took a group out to Ross River and Arltunga and then another trip out to Palm Valley. Trudy, Dad and I helped, using our one-ton Dodge truck as a supply vehicle and handling the catering.

Several months later Bond brought another bus up – this time it was a 10-passenger Studebaker which was a really big coach for those days. We continued helping on some of those tours for several years. If it was a small tour of four or five people we would take them ourselves on a contract basis for Bond. Later on he established one base in the Alice. By then Dad and I were too busy with the overland mail run to Tennant Creek and Birdum.

During one of the early Bond's Tours a geologist and his wife decided not to finish their tour to the east of Alice Springs; they wanted to go on to Mount Isa. Trudy continued as cook for Bond's Tour by herself while Dad set off with this couple via the Old Telegraph Line track to Newcastle Waters and the stock route via Anthony Lagoon towards Mount Isa.

One of Bond's first vehicles used for his tours – a President 6 Studebaker

While in Mount Isa Dad met up with Claude Nicker who wanted a lift back to the Alice. On the return trip with Claude as a travelling companion, Dad decided to travel back to Alice Springs via Lake Nash, Picton Springs, Jinka Springs, Oorabra Rockhole, Plenty River, Quartz Hill and Arltunga, using a newly opened track which would shorten the journey by about 600 miles (960 km) along rough track. The Barkly, Sandover and Plenty Highways weren't in existence then; nor was Argadargada. About 80 miles (130 km) north-east of Old Huckitta the tail shaft broke so the two men decided to walk to Old Huckitta, even though they had no idea of exactly how far away it was. All they knew was they had crossed the Sandover River earlier that day. The map they were using was fairly inaccurate and lacking in information. The track they were following had only been marked out for a few months earlier from Lake Nash to Jinka Springs. This track had been made as an access track to Lake Nash Police Station by a policeman from Arltunga, using a number of aboriginal prisoners as labourers.

It was fairly hot weather so they took a spare motor tube, cut a hole in it and filled it about three-quarters full of water (about 4 gallons or 20 litres) and hung that around their shoulders. They also took along some dry rations such as flour, tea, sugar and a bit of rice. Claude was afraid Dad might not 'make the grade' walking since his legs had always remained weak from the after-effects of polio.

To avoid the heat of the day, they walked at night. Although there was a good moon shining, the track was faint and sometimes difficult to follow because it had seldom been used. However, they walked about 35 miles (56 km) and reached Picton Springs by about ten o'clock the next morning. Picton Springs wasn't shown on their map so they still didn't know how far they had to go. They camped and rested there until late afternoon. About four o'clock they doggedly set off again, walking about another 30 miles (48 km) when they came across some cattle tracks which indicated they must be getting nearer to civilization.

When they reached Jinka Springs they camped again. There was a bit of a stockyard there and some horses (not brumbies) around which also indicated they were near to a station. Next morning they walked about another 30 miles (48 km) into Old Huckitta, one of Kidman's stations. Dad arrived there first, having lost sight of Claude quite a long way back, so a man was sent out with a spare horse to look for Claude, returning with him about two hours later. This was a bit ironic as Claude had earlier been worried about Dad not keeping up! Dad and Claude stayed at Huckitta that night, well-looked after by Maude and Billy Madrill, the manager for Kidman. The next morning they borrowed two horses and rode across to Macdonald Downs, (named for the two Chalmers sons, Mac and Donald) about 40 miles to the north-east, borrowed a T-model Ford from Mac Chalmers and drove the remaining 160 miles back to the Alice.

About a fortnight later a new tail shaft arrived on the train. Dad and I packed the old Dodge with a week's rations in the tucker box, an 8-gallon drum of water, two 8-gallon drums of petrol and set off to retrieve the Buick. We arrived about midday on the second day and put the new tail shaft in. We camped the first night at Picton Springs. Next morning we had only driven about 5 miles (8 km) when the tail shaft broke again. We hadn't realized that the torque tube, only found in the old models, in which the tail shaft was mounted was also broken, putting all the strain on the shaft itself, which it couldn't take.

It was now a case of the old Dodge towing the Buick back to the Alice over those rugged tracks where few vehicles had travelled before. That was one helluva job towing it for over 200 miles (320 km). I was driving the towing vehicle and Dad was steering the Buick. Many times when we were driving in the swirling dust he was unable to see properly and couldn't tell if he should keep on going straight or negotiate around a corner on the zig-zaggy tracks. We broke the rope quite a few times until we reached Huckitta, where we borrowed a steel rope. Even that broke a couple of times, but we finally arrived back in Alice Springs after three days.

When people went out into the bush in vehicles in those days they really had to be prepared for almost any kind of mechanical breakdowns on the dreadful roads with little access to help or replacement parts.

The Terrible Track to Jervois Range

When Jervois Range mineral deposits were first discovered about 30 miles east of Jinka Springs, it created quite a rush. The first reports equated it to a second Broken Hill. BHP (Broken Hill Proprietary) sent an eminent geologist, Doctor Aebart, up to check it out, requesting Dad to drive him out there.

A tourist party loaded on board my Federal mail truck, bound for Palm Valley

Their first two attempts to reach Jervois were disastrous. They had only travelled about 50 miles (80 km) to the middle of Bitter Springs Gorge when they pulled up so Doctor Aebart could have a look at the great rock folds there. They discovered smoke pouring out of one back wheel – a bearing had gone. Dad had only bought the Buick a few weeks earlier from Mr C.Timms, Principal of the railway construction party. Dad packed the wheel up with some leather and grease and slowly drove back home.

The two men then transferred to the one-ton Dodge and were only about 15 miles (23 km) from Jervois when the crankshaft broke. Dad got a lift back to the Alice and Doctor Aebart got a ride with someone else out to Jervois. Fortunately there were several vehicles going in and out at the time inspecting the new field. Two weeks later Dad and I went out there with a new crankshaft and towed the truck underneath a big bloodwood tree where we rigged up a block and tackle to lift the motor out and installed the new crankshaft. So we didn't get to see Jervois on that trip! However, many years later I went out to Jervois and developed it, which is another part of the story.

A party of school teachers from Melbourne travelling in air-conditioned comfort on an incomplete 'Bitzer' Mark III, leaving Hermannsburg Mission, bound for Palm Valley, 1938

CHAPTER 11
SAM IRVINE AND OTHER CHARACTERS

Sam Irvine was a pioneer of the motor mail era north of Oodnadatta in South Australia, and in the Northern Territory. In 1923 he became the first motor mail contractor for the Oodnadatta to Alice Springs and the Alice Springs to Arltunga mail runs. Sam took over from the camel mail runs which had been run by Hussein and Fred Khan, Tim and Claude Golder and Harry 'Bony Bream' Tilmouth.

Prior to becoming involved in the mail runs, Sam was a manager of a large sheep station in the north-west of South Australia. He was married with several children but his married life came to a sad ending one day when he'd intended to pleasantly surprise his wife by putting in a plug hole and drain from the tub in the bathroom. It had always been a 'drag' having to bail the water out after each bath.

Sam took his pick into the bathroom, then, finding the right spot for the drain, drove it through the tub. Leaving the pick in the tub, he went off to get some more tools to drill through the floor. He returned in time to hear a great commotion going on. His wife was thrashing the daylights out of his six year old son for putting a hole in the tub! Of course, the boy had seen the pick in the tub and climbed in to investigate, when Mother walked in on him with pick in hand.

Although Sam was a rough type of man he was also very soft-hearted, especially where children were concerned. One helluva row followed and his wife accused Sam of protecting the boy and refused to believe that Sam had made the hole. Sam finished the drain in the tub, packed up his tools, his swag and tuckerbox onto his old Reo truck and left home the next day, looking for any work to make a new start. He told how he went onto the booze for 'he couldn't remember how long' and woke up in Oodnadatta with a first-prize hangover.

He met up with Arthur Hersey, Harry Wolf, Gene French, Edgar Horwood, Gus Brandt and Ivor Windelman, Phil Windle and Billy McCoy. These men were some of the pioneers of motor transport north of Oodnadatta from 1923 to 1928 with their one and a half ton Reos, Dodges, Willys Knight and T-20 GMC trucks and a 3-ton Willys Knight, owned by Harry Wolf which was thought to be very big at the time.

Sam and his trucks – first a 1923 Reo, then in 1927 a T-20 GMC, in 1929 a T-30 GMC and in 1936 a 2-ton Federal – were the subject of many a story of perseverance, courage and in fact, heroism at times, that have been related down through the years by old-timers of the region and mentioned in several books. A humorous incident took place at Newcastle Waters River Crossing.

Irvine charged through the water and mud at high speed, staggering half-way up the other bank. The motor stalled at that point. Sam and his off-sider, George Nichols, decided to have lunch of a tin of corned beef and a bottle of beer while the motor dried out. Sam then took out the crank handle and cranked away for half an hour with the throttle wide open. The engine eventually started with a roar, a loud bang and a rattle. It promptly stopped again with one con rod sticking out of the crank case. Sam looked at it, then at the crank handle in his hand. He said, "Well, we won't need this f------thing again!" and flung it as far as he could into the river.

After another beer he looked over the damage and decided to patch up the hole in the crank case with leather, a few screws, a piece of oil drum and flour dough. He then tossed the piston and bent con rod into the tool box. "Right!" he said, "She's all set. Let's go! Where's the f------ crank handle?"

George Nichols replied, "Where do you think it is? In the f------ river where you threw it, of course!"

Three hours later, after a much-needed wash and a cooling of tempers, they found it and carried on with five cylinders functioning.

As the railway line extended northwards from Oodnadatta, Sam's mail services were adapted to changing needs of the area and he relinquished the Arltunga run (which we tendered for) and he continued taking the mails from Alice Springs to Tennant Creek with later extensions to Newcastle Waters and Birdum, also later taken over by Dad and me. When Sam discontinued the mail runs he worked for the government operating a road grader for many years until he died in Alice Springs in 1959. He was buried in the Alice Springs Memorial Cemetery.

Some More Bush Characters along the Mail Run

Early one Sunday morning in 1933 a knock came at our door. It was someone with a message from Sam Irvine, asking if I could do the mail run up to Birdum for him because he was too sick. Sam was a heavy drinker and after imbibing a bit too much, if I remember rightly, an ulcer had erupted.

At that time I had built the old Dodge 4 truck up to a fairly good standard. Doug Adamson was the Postmaster at Alice Springs then, from whom I had to obtain a special permit to do the run, although seeing how I was already doing the eastern mail run, it wasn't any trouble.

That was my first trip right up north going through Tennant Creek which was quite a new experience and an eye-opener for me at the age of eighteen. I met many interesting characters along the way, all of whom wanted to know where Sam was and whether I'd brought their booze along. Of course I didn't know anything about the booze! Apparently Sam had regular orders for a bottle of rum for this fellow and a bottle of brandy or a case of brandy for somebody else and so on. There were quite a few disappointed old cronies of his on that trip!

Rough 'Justice' at Tennant Creek

Tennant Creek was in its infancy then. In even earlier days, Joe Kilgariff from the Stuart Arms Hotel in Alice Springs took his Chev 4 up towards Tennant Creek loaded up with beer, sheets of iron, timber and cement, intending to build a hotel somewhere near the old Telegraph Station. About 8 miles (13 km) south of the Telegraph Station his vehicle became thoroughly bogged. I think an axle broke which finished it off, so he unloaded everything and set up camp right there and started the new pub on that spot! That is how the town of Tennant Creek happened to be built where it is. Joe Kilgariff started building his pub, followed by Alex Scott and Jack Noble who came across from the West. Jack bought Joe out after striking rich gold at what is now called 'Noble's Nob', so the town grew around the pub.

When I first went up there, Scott's Hotel was in the process of being built and was almost completed. The other hotel, known as 'The Goldfield' or McMahon's, had only just commenced trading. Tennant Creek was quite a 'wild and woolly' place in those days, including a few gun battles every now and then.

I was told how there was a bit of a court case once when Al McDonald was peppered with gun shot. Al was a Maori 'pug' who had been quite a fighter in his day. However, he used his boxing ability to make money in a different manner in Tennant Creek. He'd offer to finance 'battlers' who had a good prospect and lease to mine a gold or wolfram show somewhere. He made an agreement that he'd collect the returns and take his cut for staking them, giving the owner whatever was left over.

It was said that one of his ways of payment was to take an unsuspecting miner, whom he had staked, into the pub and buy him plenty of booze until the poor bloke was nearly rotten. When the miner would say, "Come on, give us me money! We want to get our stores tomorrow and get back out to the bush!" Al would make a big 'song and dance' about it (so everyone in the pub could hear), pull the money out, counting out £200 or £300, whatever the amount was to be paid. He'd buy the fellow a few more beers and before long his victim would be staggering out to the toilet which was down in the back yard. Al would follow the chap out in the dark, drop him with a punch and take the money back, just leaving the poor fellow a few pounds in his pocket. His victim being knocked out and gone to sleep, wouldn't remember anything about what had happened.

When he'd finally awake the next morning he'd find most of his money was gone and wouldn't even be sure whether he had been paid. He'd then go and ask Al when he could get his money. He'd be told, "I paid you last night and I have witnesses." Then Al would finance him again and call him "a bloody old nuisance!" Next time the same thing would happen. Eventually a bloke who had his suspicions pretended he was drunk and McDonald allegedly tried to cut the same caper. I don't quite know the full strength of the story, but the incident ended up in a shooting, with Al McDonald picking little lumps of lead out of himself for quite a long while afterwards. Scores were settled in that fashion! There were tales of several other minor shootings at Scott's Hotel, but no fatal ones that I can remember hearing about.

Fitzy, the Fair Cop

Police Sergeant Fitzgerald at Tennant Creek was well-liked, well-respected and known to all as 'Fitzy'. Fitzy resembled his bulldog, known to everybody as 'Sarge', both of whose faces bore record of their fighting careers. Fitzy controlled law and order in the town of about around 500 or more miners, almost single-handedly. He was later accused of being too lenient by his superiors when they found he didn't keep a crime sheet for minor offences and was eventually transferred from there.

Fitzy's method of justice suited the population and was appreciated as being a fair way of dealing with the problems. Anyone who started a fight or caused trouble was taken down to the 'cooler' which was a little tin shed, and left there for the night, or given a good hiding if warranted and next morning they were released without being charged. If they were broke he'd give them a quid for a reviver and told them to get out of town.

Another example of Fitzy's justice, which probably didn't please his superiors, was his policing of hotel trading hours on Sundays. Drinking on Sundays was restricted by law to travellers and guests staying at the hotels. Thus the miners who worked hard all week out at their diggings could only enjoy a drink on the weekend on Friday nights and Saturdays. Fitzy, being a fair cop, with an understanding of a miner's thirst, allowed the pub to open for one hour – between eleven and twelve o'clock – and under his supervision the suffering line filed in to enjoy an extra thirst-quenching session. Fitzy policed the exits to make sure that no alcohol was taken off the premises. He was also fair to the licencees in that he rostered which pub would be open on Sundays for the hourly reviver session.

Meeting Others along the Track

I proceeded on to the Telegraph Station at Tennant Creek, about 8 miles (13 km) further north where the water supply was. All water was carted to the Tennant Creek township until about 1960. 'Cock' Martin was in charge of the Telegraph Station at the time and gave me directions for the trip north and told me a bit of information about the track.

My next stop on the mail run was Banka Banka Station. The owner wasn't at home at the time and there were only a few aborignes there. I left the mailbag and went on to Helen Springs where the Bohnings lived – Mrs Bohning, old Jack and their two sons, although the two sons were not home at the time. I was invited in for a cup of tea and a meal. Old Jack said, "Come around this way into the kitchen." He shooed away chooks, pigs and goats along the way. Inside the kitchen there were dogs, cats, goats, chooks – everything was topsy-turvy! However, although it was a real old 'hill-billy' type farm, you couldn't find kinder people than this old couple.

Old Jack and Mrs Bohning were drovers before they settled at Helen Springs. Mrs Bohning told me how she used to bring the wagon on and do the cooking. I think she gave birth to two or three of their children along the way. Her husband and the drovers would take the cattle on early in the mornings. One day, after the men had all left, her baby arrived about two weeks early. After the baby's birth she packed everything up and continued on with the wagon just as if it was a routine event. Two of their boys settled in Alice Springs.

From there to Powell Creek the track was very rough for about another 30 miles (48 km). There were some characters there too! Powell Creek was another repeater station on the telegraph line. Old Wallaby (Wally) Holtz, who was about 70 years old at that time, was the telegraphist. He had a bulbous, deformed-looking nose which made his appearance almost ugly, but he was a really pleasant, well-educated person. He lived along the creek with an aboriginal woman, Rosie, some distance away from the Telegraph Station. Apparently the old rule was that employees were not permitted to have aborigines in the station quarters, so he had to move out if he wanted to live with Rosie which he did quite happily for about 30 years.

The boss there, 'Bro' Hall, wasn't quite as pleasant. He was officious in manner and a 'dobber'. If the mailman was an hour early or an hour late he would get on the 'tick-tack' (morse key) and report it to headquarters in Adelaide. If old Wally was on duty he told me he would 'tick-tack' the message but wouldn't switch it through!

From Powell Creek the track continued on up past Lake Woods to Newcastle Waters. It was a great sight and quite a surprise coming over the hills to see Lake Woods in the distance. The telegraph line and the track ran right alongside of the lake, about 50 metres from it. On this first trip I made up there the lake was what they called 'normally full'.

At Newcastle Waters I met Max Schauber, an old German publican, who always had a grumble about something, but was really very soft-hearted and would do anything for you if you didn't take any notice of his growling and grumbling. There I also met Jack Althouse, Sam Irvine's pack horse man. Jack took the mail on from there because the swamps further north were filled with water. I waited at Newcastle Waters for a few days until Althouse returned with the south-bound mail. I stayed with Jock and Mick Jones, who had a store and bakery in the little township. Jock also did fencing, yard building and aerodrome maintenance in the district. Norman Stacey brought the mail from Darwin to the Birdum railhead and met Althouse halfway to exchange mail bags.

Cooling the Beer

I had just delivered a ton of beer in five dozen wooden crates to Max Schauber's pub at Newcastle Waters. The bottles in those days were packed in straw sheaths. Since the pub was right out of beer Max immediately unloaded the beer with plenty of willing helpers such as drovers and so forth, stacked the beer in its straw-covered bottles on some cyclone wire racks under the bar counter and threw some water over them which was his normal practice for cooling the beer.

Of course the beer was still warm, coming straight from the truck – it would have taken several hours to become cool enough to enjoy. However, the customers were all eager to have their beer so each one was handed a bottle. When they tested it, Jack Althouse complained to Max, "This bloody beer is as warm as piss! Why don't you get a refrigerator?"

Max walked out of the bar to the store section to attend another customer, grumbling, "What else can you bloody expect up here in this heat!"

While Max was away Althouse went into the kitchen, grabbed a handful of saucers and put one in front of each bar customer. They each poured some of their beer into the saucer and waited for Max to come back. As he walked in they all started to blow on their beer to cool it down! Max didn't take the joke very kindly and said in his thick German accent, "You bloody convict bastards! Always complaining! Why don't you go back to where you come from, but then I suppose you'd complain it was too cold!"

On another stinking hot day the Bathan brothers, nicknamed the 'Bullwaddy Boys', so-named after the impenetrable Bullwaddy scrub on their property, rode in on horseback and got a bright idea of how to stir up Max a bit further and cool things off in the pub. The whole pub was built of steel and corrugated iron with the walls raised about 8 inches (20cm) from the concrete floor, which allowed for ventilation and for hosing it out for cleaning purposes.

The Bullwaddy Boys gave their horses a drink of water from a 44-gallon drum which was kept on the verandah for general use, then walked into the bar. Some city travellers were inside and one of them was complaining bitterly about how terribly hot the place was, making quite a fuss and saying the hotel should have better cooling and ventilation and so forth. The Bullwaddy Boys cheerfully said, "We can fix that for you – no worries!"

They walked outside and upended the 44-gallon drum of water, which sloshed through under the walls into the bar and flooded the feet and shoes of the city slickers. The Bullwaddy Boys came back in with big grins on their faces and asked, "How's that, mate? Any better?"

A Couple More Characters

My trip back was quite uneventful until I reached Tennant Creek. There I met Eric Miller and Jack Hicks, two new 'blow-ins' from Victoria who'd come up for the gold rush and obtained the contract for the Tennant Creek weekly mail. Eric had driven up there in an old Studebaker truck. It was rumoured up there that Miller and Hicks had been involved in the 1932 possum skin murder mystery in Victoria. A semi-trailer load of possum skins in wool bales disappeared and the driver was found dead. The semi-trailer was found abandoned months later, but the possum skins, which were quite valuable in those days, were never found. It was thought that they had ducked off to the Territory to let things cool off a bit. At any rate, Miller had the mail contract for the Tennant weekly. They only just made it into Tennant Creek when Miller's old bus broke down, so they asked me if I would take the return mail back to Alice Springs for them.

I no sooner arrived home in Alice Springs when they asked me if I would also do their next mail run up, which I did, so I had two trips up to Tennant Creek within two weeks. I had to wait in Tennant Creek for three days then, for the return mail. I had time for a bit of a look around. The El Dorado mine was just starting to produce and several other mines were being opened up. Men from all over Australia swooped into Tennant Creek. Some were honest men and 'go-getters,' some were legitimate miners and some were simply crooks looking for someone to 'fleece'.

There were two hop-beer shops which produced a real 'tangle-foot' brew, containing all sorts of concoctions to bring the alcohol content up. It sold quite well and one brewer claimed that there was 'a fight in every bottle or you get your money back'. I quite believed it too from the racket that went on over the weekend.

Jim Maloney and his wife, who had come across from Queensland, ran one of the better-run shops. Jim was also the local bookie. He was quite a heavy fellow with one wooden leg. A lot of the battlers up there were 'on tick'. If they had a bit of a claim and could show that they were finding a bit of gold, Old Jim would back them. It was the same with food supplies. Harold Williams opened up a store there and was also very good to the battlers. He would always give them credit if they looked like a 'right-kind-of-a-bloke'. He wouldn't give credit for any grog and if he knew they were boozers he would think twice before supplying them.

Con Perry was a Russian who set up a butcher shop there. In later years he took up Phillip Creek Station, which was a small holding about 30 miles (48 km) north of Tennant Creek. In those days the stations didn't have any fences so he bought 30 or 40 head of cattle from the neighbouring stations for the butcher shop and added any strays which walked south to Philip Creek, which had large water holes. Since there was no water beyond that point the cattle would stay there. By purchasing a small number of cattle from each station bordering his northern boundary Con 'acquired' an almost endless supply of cattle with those brands. The locals reckoned he must have multiplied those bullocks by quite a number and developed a profitable business.

Those were a few of the characters I met on that first experience of taking on the relief mail runs up to Birdum and back for Irvine and Miller in 1933.

Bob Buck, Gus Brandt and Sam Irvine, 1952 (Photo courtesy Judy Robinson).

Chapter 12
Anthropologists to Mount Liebig

In January, 1934 I took on the job of driving an anthropological party from the Adelaide University from Alice Springs to the western side of Mount Liebig. They wanted to obtain accurate data about the aborigines who were living out there, still relatively untouched by civilization. They intended to study their lifestyle and culture and record some of their corroboree songs. We set off using my old Dodge one-tonner. There were about seven people in the party including Professor Davies and Professor Cleland.

We stacked all their equipment and stores on board with most of the members of the expedition also perched on top of the load. Professor Davies drove a Baby Austin with one passenger riding with him. About half-way out the little Austin broke down – its magneto 'packed up', so we stacked all Davies' gear on top of the Dodge, plus the extra 'bods' and trundled off leaving the Austin at the side of the track.

We were following a very faint, seldom-used bush track and had just passed Haasts Bluff on the northern side when I noticed that the steering of the Dodge felt rather strange. I pulled up to have a look and discovered that one front wheel bearing had collapsed. The inner bearing had fallen to pieces and was lost somewhere along the road.

Since it was about lunch time I said, "While I think about this, you all have some lunch." The flies were something terrible and it was stinking hot. The 'greenhorns' aboard from Adelaide University were absolutely horrified at our predicament. Some of them were actually shaking with fright because we were in such remote country with only miles and miles of mulga scrub around us. We had no radio or other means of communication and the nearest settlement at the time was Hamilton Downs Station, about 50 miles (80 km) behind us and 20 miles (32 km) off the track. It would have been a mighty long trek if we had to walk back there! Nobody was likely to come out looking for us yet because we weren't due back in Alice Springs for a week or two.

While they were opening up the 'tucker box' I jacked up the front of the truck and pulled the wheel off. The outer bearing was still okay but there was a gap of about half an inch (1 cm) all round the back. On board I carried a big coil of 10-gauge galvanised wire and a large tin of grease. I decided to wrap a layer of the galvanised wire on the inside to fill the gap where the bearing had been, put some grease on and put the wheel and outer race back on. It seemed quite firm and was quite smooth in the old outer bearing so I said, "Right! I'll have a bit of lunch and we can get going again!"

Apparently they had expected me to turn the truck around and return to Alice Springs. Professor Davies thought we might be able to go back and fix the magneto on his Baby Austin (50 miles or 80 km back) and save a lot of walking, but I was ready to head straight on out to our original destination. They were quite horrified when I suggested we'd go on. There was still about another 60 miles (97 km) to go before we'd reach the proposed base camp, but it was 100 miles (160 km) to go back to Alice Springs.

There were strong objections voiced and I was accused of being irresponsible in attempting to go further. I persuaded them to give it a try by saying, "Righto, let's give it a go and see how long the wire lasts." We tootled along with the wire working quite well. About every 15 miles (24 km) I had to jack the ute up and put some more wire and grease on. By this time the group were gaining more confidence in me, so we proceeded in this manner until we reached Mount Liebig rockhole and made camp.

There were no aborigines in that vicinity so we decided to go around the end of the range to the southern side where there was a permanent water hole about another 30 miles (68 km) on. Before taking the truck to the second water hole we decided to see if we could walk across over the top of the Mount Liebig Range and see if the aboriginal tribe was actually camped there. This was a distance of about 8 miles (13 km) 'as the crow flies'. Two other fellows and I left early in the morning and after climbing over the top of the range, arrived at the aboriginal campsite just before midday.

The camp was deserted, but there was evidence that the tribe were still living there. By sundown a number of them returned from their hunting. An interpreter had been sent out earlier from the Hermannsburg Mission to let them know we'd be coming. The thirty or so aborigines living there received us in quite friendly fashion. This group had been in some contact with patrols from the Mission who had carried out surveys to see how many tribes were in existence in the surrounding areas.

Aborigines near Mount Liebig, 1934 (Photo taken by Pastor F. Albrecht, courtesy Helene Burns)

The rest of the anthropological party, having taken the longer route to the east through a gap in the range, arrived in dribs and drabs during the afternoon. It was quite warm weather so we camped out under the stars with no swags and waited for the last stragglers to arrive. Two men were still missing by nightfall, so another man and I walked back to search for them. We found them suffering from cramps and unable to walk. Some others in our party had also suffered from cramps due to loss of fluids and being unaccustomed to the extreme heat. We brought them back to the campsite at about ten o'clock that night. They had seen the fires we had lit on the hill to guide them to where we were, but the bad cramps in their legs made it impossible for them to walk until it cooled off.

Next morning Professor Davies looked around the aboriginal camp for some salt and in a humpy which belonged to one of the mission boys, he found a bag containing some crusty red lumps of salt. The red tinge was from blood where it had been used for salting kangaroo meat. He dished out a portion of salt to those who had suffered from cramps the day before.

While the main party remained at the campsite, another man and I walked back over the range to where we'd left the truck, packed up the gear and cut a track a further 30 miles (48 km) over some very rough terrain, through gullies and scrub around the western end of Mount Liebig Range to bring the supplies and equipment to the expedition members. They were extremely happy to see us since they'd had nothing to eat all day other than some kangaroo meat offered to them by the aborigines which only two of the group accepted and ate. I remained with them for another night and we unloaded all their gear the next day, which was my nineteenth birthday.

Some of the expedition members became quite worried as I prepared to leave for Alice Springs on my own in my truck with its 'crook' front wheel, my tin of grease and coil of wire! They were in a bit of a quandary about what to do in case I didn't make it back safely. If I only got halfway back, started walking and perished in the desert they'd be stuck there for goodness knows how long! They debated whether or not to send somebody with me. I reasoned, "Well, if we have trouble and have to walk out, I'm used to the 'Bush' and the climate, but you're not. You're more likely to get cramps and perish in the heat. I'd rather go alone."

They had to accept that but they couldn't know if I arrived back to the Alice or not until three weeks later when I was due to pick them up for the return trip. Anyway, I drove off on my own and kept up the same routine of putting in wire and grease until I reached the Baby Austin. I decided to load it onto the truck and take it back to the Alice to fix the magneto. I found a creek with a steep bank on one side. I tied the Austin on a short rope and towed it to the high side of the creek with the hand-brake partly on, so that it would follow without bumping into the back of the truck. Next I backed the truck into the creek, put the Austin into first gear and slowly wound it onto the truck by using its crank handle. After tying it down I drove on to Dashwood Creek which had a wide sandy crossing.

After working my way across the sandy creek I was hot and 'buggered', thirsty and hungry. I went down to the 'soak' which we had previously dug in the middle of the creek and sat in it to cool off. I lay in the water for about half an hour, which felt great! I then took a tin of green peas out of my tucker box and quenched my thirst by drinking the juice before eating the peas. I drove on a few more miles to get away from the mozzies in the creek and made camp, boiled the billy, cooked up a little damper, had a feed and rolled out my swag on the flat.

Some time during the bright, moonlit night I was awoken by a great "THUMP! THUMP! THUMP!" Just as I opened my eyes a huge form flashed over me, which gave me a helluva fright. As I woke with a start, I realised it was a big kangaroo, but my heart was still pounding! I expect the kangaroo's heart was thumping too, not expecting to find a human being in a swag lying across his pad, way out there in the mulga scrub!

Everything continued well until I was only about 35 miles (56 km) away from Alice Springs, near Painters Spring, when I ran out of wire. I had to use my 'thinker' again considering what to do as I looked through all my bits and pieces. In my tool box I had a pinion bearing which was about 1/8th of an inch too small in the centre hole. I also had a file which wasn't in the best of nick, but I sat down for about three hours filing down the stub axle so that the bearing would fit. It was a bit loose on the outside, but that didn't matter. Eventually I attached it, put the wheel back on and drove the rest of the way home. Back there I mended the Baby Austin's magneto by fitting a coil onto it and converting the magneto to battery ignition.

Some aborigines out at Mount Liebig (Photo taken by Pastor F. Albrecht, courtesy Helene Burns)

Two and a half weeks later I set off again on my own to pick them all up. I don't think I've ever received a happier greeting! They had no way of knowing if I'd even returned safely to Alice Springs after leaving them. They were down to their last provisions which I'd anticipated, so I'd brought out extra flour, tea, sugar and some meat over and above my ordinary emergency rations.

The Professors had recorded one of the main corroboree songs, sung by Wunjawara the chief, using a cylinder phonograph. They played it back to him. When it was finished he turned around to the young men of his tribe and said in disgust, "I bin tryin' to teach you dumb lot this corroboree a long time. This white fella only heard it once and he didn' make one mistake!"

Next time the anthropologists came up and asked me to take them out on their second trip they brought a radio with them, thinking they'd be safer. However, the transmitter didn't work and they didn't know how to make contact with it. They could hear the base but that was all, so it wasn't much use to them.

Some aborigines listening to a gramophone, photographed by Pastor F. Albrecht
(Photo courtesy Helene Burns)

Chapter 13
Stud Rams to Kerr's

Around 1934 the stock agent in Alice Springs asked Dad and me to take a load of ten young stud rams out to Alec Kerr's sheep station, Old Delny, 130 miles north-east of the Alice (the same fellow who'd bailed me up with the rifle on my very first mail contract).

We made up a crate around the back of the old Dodge 4 truck (which had done close to 60,000 miles by that time) and set off one winter morning. We had travelled to within 15 miles (23 km) of Delny when the stub axle on the right hand side broke and the front wheel fell off, bringing us to a sudden un-scheduled stop! The axle had simply snapped off due to metal fatigue. Using bushes and sticks we built a bit of a yard up for the rams, unloaded them and made camp for the night.

At daybreak, after breakfast, I left Dad there with the rams and started off to jig-jog-run and walk to Alcoota Station, about 25 miles (41 km) back, which we'd passed through the day before. I was hurrying because Alf Turner, the owner of Alcoota Station, had told us they'd be leaving early in the morning for mustering, and I wanted to get there before they left.

As I trotted along in the frosty weather I thought I heard a motor, so I stopped and listened. I couldn't hear anything so I resumed my walking until I thought I heard a motor again. I stopped, listened and waited, but heard nothing more so I went on again, thinking it must have been my ears playing tricks on me.

Even though it was only 8:30 a.m. when I arrived at Alcoota, the mustering group had already left, having also made an early start. Mrs Turner apologized, "I'm sorry, but all the horses have gone. There is only one old mare left in the paddock. If you think she'll make the journey to Bushy Park (which was another 45 miles or 73 km further back) I'll get you a saddle and you can take her." Which I did.

I pushed that poor old mare along all day, sometimes getting off to walk beside her for a while. I finally arrived at Bushy Park Station about six o'clock that evening, turned the mare loose at the well to find her own way back home, and walked into the station carrying the saddle and bridle.

Jimmy Bird was all packed up, just about ready to go off to a neighbouring station. He stated, "You're bloody lucky I'm still here since you've let the horse go! I was just about to leave!" Another half an hour and he would have been gone! Anyhow, he agreed to stay at home for a couple more days and let me borrow his old Chev truck to go back to Alice Springs and get a replacement axle. I drove the 60 miles (97 km) into town, arriving there about 10 o'clock that night. I was feeling very stiff and sore, being unaccustomed to horse riding, plus having walked quite a long distance since that morning.

Early the next morning I took a stub axle from another old wrecked vehicle at home, filled up the borrowed truck with petrol and returned to where I had left Dad. Lo and behold, the truck and Dad had disappeared, along with the rams! I could see from his tracks where Dad had apparently turned back and come part of the way back to try and catch up to me walking.

What actually happened was Dad had decided to make a wooden stub axle out of a large mulga fork stick by tapering the thick part to fit through the wheel where the bearings had been and, using the brace and bit which we carried in the toolbox, bored a hole through the protruding end. He then pinned it in place with a bolt, leaving the other two ends resting on the spring. He used another bolt going through behind the broken stub axle to keep the fork stick in place.

He then turned around to follow me, which accounted for the motor sounds I had first heard. While he was following me the left side stub axle broke, which shows how even the steel must have been in that old vehicle. He made up another wooden axle, turned back to pick up the rams from the make-shift yard and set off towards Kerr's station.

When he encountered another creek one of the wooden axles broke, so he jacked the truck up and patiently fashioned yet another mulga stub axle. However, the creek bank was too steep and the wooden axles couldn't take the strain. He unloaded the rams once again and started shepherding them towards Kerr's station, leaving the truck in the creek. Dad still limped fairly badly from the effects of his broken leg and was using a walking stick. I caught up with him about 3 miles (5 km) from our destination. We left the rams where they were and drove on to let Kerr know what had happened. He sent two aboriginal workers back with me to collect the rams.

Dad had left the Dodge in the creek with two wooden axles, one broken and the other still holding. Of course I had only brought one replacement stub axle out and now we needed two! We decided to go a further 22 miles (35 km) to Bill Petrick's place at Mount Swan, knowing he had some old trucks there.

In the meantime a thunderstorm and rain had flooded part of the road to Mount Swan. It was only a track which hadn't been used much, as Bill was only just settling into his new station. It took us about five hours to travel 10 miles (16 km) bogging along. We left the thunderstorm behind and arrived at Bill and Elsa's place where we stayed for three days waiting for the road to dry up.

When we compared the stub axle from Bill's Chev truck with the Dodge's they were totally different, with a thinner kingpin and a few other differences. However, we decided we could make it work by modifying it, providing we took the wheel and tyre as well, together with a few bolts and wire. We drove back to the disabled Dodge, mended it and returned to the Alice with one Dodge wheel and one Chev wheel, also returning Jimmy Bird's borrowed Chev at Bushy Park. The Dodge looked a bit strange with one wheel bigger than the others. That had turned out to be quite an arduous trip, taking seven days instead of the two days on which we had originally planned. For many years I kept that old wooden stub axle as a souvenir of the trip. In the end it disappeared; one frosty morning it probably found its way into a fire bucket to keep somebody warm at the back of the workshop!

Dad had actually travelled about 15 miles (23 km) with one wooden stub axle and when the other one broke he travelled a further 10 miles (16 km) with two wooden stub axles, which is quite a remarkable feat. It is an example of Dad's perseverance and ingenuity which also rubbed off onto me by way of example from very early in my life. It shows what a person can do when 'in a spot'.

Chapter 14
Foy's Expedition to the Western Desert

Late in 1935 Vic Foy, a millionaire globe-trotter, brother of Mark Foy from Foy's Limited in Sydney, inquired from the government offices in Alice Springs if there was somebody 'up there' who could organise an expedition out to Lasseter's country. Vic suffered badly from asthma and every winter he'd find somewhere in the world with a dry climate to which he could travel. He wanted to go out to Lasseter's country and to Ayers Rock, to make a film. He was informed that I'd taken out anthropological parties into the west and they recommended that he contact me, which he did.

It was then that I built my first 'Bitzer' using assorted parts or 'bits and pieces' from various vehicles, which is described more fully in a later chapter. It was towards the end of May in 1936 and I was running late in getting the 'Mulga Express' Mark I finished. The expedition party had already arrived up by train on Saturday, bringing their one-ton Chev truck and we were due to leave on the following Monday.

Kurt with 'Bitzer' Mulga Express Mark I and Paul Johns, 1936

On Saturday I gave 'Bitzer' her first trial run around, picking up fuel, water and stores. We carried 100 gallons of petrol and 100 gallons of water. Two weeks earlier I'd sent Ben Nicker and an off-sider ahead to select a campsite beside water on the Docker River near the Western Australian border. They took out six camels loaded with a ton of petrol and stores, plus water for himself and his off-sider.

The expedition consisted of six men and one woman (Mrs Foy), and their 12 year old son. Early Monday morning we headed more or less in a direct line towards Ayers Rock via Owen Springs and Middleton Ponds. There was a track as far as Middleton Ponds, about 100 miles (160 km) out. For the remaining 120 miles (192 km) to the Rock we picked out our own track over desert, sandhills and scrub.

The photographer/signwriter in Foy's party christened Bitzer Mark I "The Mulga Express" after experiencing the way it crashed through the mulga scrub. While he was painting "Foy's Expedition 1936" on the bonnet he decided to add "The Mulga Express" on the side panel.

We had quite a good run until we were about 40 miles (64 km) from Ayers Rock when all the rivets sheared off the crown wheel in my truck's differential. The party members almost immediately went into a panic, wondering what would happen now. I said, "Well, let's have lunch now while I pull it out and see what I can do." Thinking it might have a broken axle, I took the shovel out and dug a hole under the back of the truck since we were sunk well down in the sand. When I pulled the back off the differential and found the trouble, I knocked all the broken rivets out and found half a dozen short bolts in my tool box. I pirated another six out of the chassis here and there, cut them to size, heated them in the fire and riveted the crown wheel back together. And that did the job.

Three hours later we were driving along again. We had no more trouble after that on the entire round trip of nearly 1,000 miles (1,600 km) except for one puncture. In fact, those rivets stayed in there for quite a few years until "Bitzer Mulga Express Mark I" was scrapped.

When we arrived at Ayers Rock we camped there for two nights and climbed the Rock on 28th May, 1936. On top of the Rock there was a Vesta metal matchbox with a couple of names scratched into it and a bottle with three or four names in it of people who'd previously climbed the Rock. My name became the fifth name on the list. When all our names were added to the list the number of names in the bottle had reached twelve.

Ayers Rock

The aborigines living in the area would never camp close to Ayers Rock because of sudden winds which could occur during the night. It might be a dead calm night until maybe ten o'clock at night or two o'clock in the morning, when there'd be a sudden whirlwind of high velocity. However, the people in our expedition didn't think it would be a problem and erected their tents anyway, right up against the Rock.

The second night it happened! Suddenly the tents began violently flapping and tables were tipped over. Just as suddenly it became dead calm again, as if somebody had given a giant puff and then disappeared. The aboriginal legend was that the 'Kadaitja' or spirit didn't like anybody camping close to the Rock, so he'd try and move them away. The winds were actually caused by the thermals – hot air rising from the hot rock.

The following day we travelled on past the Olgas to Lasseter's Cave and took several days to travel further to the Ruined Ramparts, all of which was fabulous scenery. We finally arrived at the Docker River soak and made camp, did a bit of prospecting for a few days and some filming was carried out. Unfortunately, when Foy returned to Sydney it was discovered that the film had been ruined due to the extreme heat. The early films couldn't stand high temperatures. Although that aspect of the trip was a failure, Vic Foy was quite pleased just to say that he'd been out into Lasseter country.

At that time there were still many aborigines in that area who had seldom or never been in contact with white people. We almost had a catastrophe. Ben Nicker put out some baits for dingoes and next morning when we went around to collect them we saw tracks where the aborigines had picked up the dead animals. We hurried to find them before they could eat any, but when we arrived at the Aboriginal camp they had already cooked the dingoes and were eating them. The poison must still have been in the stomachs of the dingoes because it didn't hurt any of the aborigines.We stayed out there at Docker River for about a week. We were away for a total of four weeks, including all the travelling. It was a really interesting trip through some of the marvellous scenery of Central Australia.

When I look back on it now I am astonished at the naive way in which I took this trip on, but I was young and enthusiastic and had plenty of experience in desert conditions. I had no hesitation about heading out into the desert country and finding my way out there and back. Nowadays, if I was planning to do the same trip I'd make a lot more preparation and wouldn't dream of taking a party like that out without a two-way radio. Of course, back then I didn't even think of a radio because they weren't readily available. Also, as far as my truck was concerned, I hadn't even had time to test it out or take any decent runs to check it out and make sure everything was okay. I suppose I had what you could call confidence in what I had built. I was young and daring, but ignorant of how city folk or strangers to the bush could quickly become panic-stricken when something went wrong in those remote areas.

I met up with the Foys again in later years and they were still talking about how much they had enjoyed that trip and seeing the lovely country out west, even though it was pretty rough and no home comforts. We did put up a tent for them!

Another meeting with them occurred at Daly Waters. I was still in my swag camped near the pub, when some people arrived fairly early in the morning. I heard a familiar voice saying, "That's Kurt Johannsen's name on that truck!" A gruff old voice answered, "Oh, he wouldn't be up here." Next thing I heard Mrs Foy's voice again so I got out of bed and surprised them by appearing! They were quite excited to meet up with me again. I only met with them once again several years later in Sydney when I stayed for a couple of days at their home.

A loaded Mulga Express Mark III, on the way to Birdum with the mail

Chapter 15
A Holiday Down South to the 'Big Smoke'

Late in 1936 I decided to take a holiday down to the 'Big Smoke' in Adelaide and Melbourne. Work had been going pretty well and I'd saved a bit of money from Foy's expedition, plus the Underdowns owed me money for carting sand, building materials and making general deliveries for the hotel they were building. A couple of other people also wanted to come down south with me, sharing part of the costs for petrol.

I needed to collect the money owed to me by Ly and Daisy Underdown. Daisy was known by many as 'The Black Widow' because she was always dressed in black and could always be found in her office from morning until night; the office was her centre of control for her web of financial transactions. In those hard times during the Depression, many people used to book up purchases and transactions rather than paying cash. I did too and often, on hot days at knock-off time, I'd taken the two half-caste lads who worked for me into the bar at Underdowns where I'd 'shout' them a couple of beers which was entered onto the slate or day book. I knew I hadn't spent very much and figured that I was owed about £150, after expenses.

For a couple of weeks I kept asking Mrs Underdown for her account of my expenses, but she wouldn't give it to me until I presented my account. She always put me off and said, "I haven't got it ready, but you bring your account around and I'll pay you." Dick Turpin, who used to work for her, had warned me about some of her tricks and told me not to give her my account until she gave me hers. He explained that when items were booked up for bar trade, she'd put down say, £1/10/- for something on one line of the day book, like other business people did, and afterwards she'd fill in the rest of the line with extra amounts so that an account might end up three or four times more than the correct amount when it was presented. Dick said her idea was that after someone had a few drinks he wouldn't remember how much he'd booked up.

Eventually we were all packed and ready to leave on our holiday. We pulled up in front of the pub and I went in to settle the account and receive the money owed to me for the carting I'd done for Daisy. I was dismayed to find that she had contra'd out most of my account down to £38 which I had to accept because we were packed up and everyone waiting outside and ready to go. This was a great blow to me because I knew I didn't owe her as much as she had deducted. However, I didn't have a leg to stand on because it was her word against mine and I knew she would simply bring out the Day Book with all the extra entries in it. I only had about £20 in the bank, which gave me just over £50 to go away with on the holiday.

There were six of us setting off on the trip – my sister Trudy, Ken and Claire Harris from Tanami and two other lads. We had quite a good trip down, travelling by day and night, taking 55 hours. The old track down to Adelaide was very rough, running from station to station or from bore to bore, zig-zagging about. We went via The Pines, by-passing Kingoonya, down to Port Augusta. It was a gate-opening trip (46 of them) through the sheep country.

On the way down to Adelaide one of the lads had some Vesta matches loose in his pocket, and

Claire and Ken Harris, the two lads, Harry McConville and Kurt after putting out the fire from the burning tyres

when we were about 100 miles from Adelaide, he took off his dirty shirt and stuffed it under the spare tyres on the back of the truck. Friction must have set off the matches and next thing we knew our tyres were well and truly alight and half the tray was also burnt through! Harry McConville from a nearby farm came racing down to help us put the fire out before it could burn his crops. That was the end of my spare tyres, which was a real problem later.

I needed some new tyres very badly and wanted to go across to Melbourne to collect some tyres from McCartney's Transport which they had previously borrowed from me up in the Territory. McCartney's had had a contract up North, using three Buzing Nag German trucks, to cart all the equipment for building The Battery at El Dorado Mine at Tennant Creek. When they left they'd promised they would send me some money for the tyres, but they hadn't up until that time.

In Adelaide we also met up with Paul Johns and his mate, Jack Hicks, who also wanted to travel across to Melbourne with Trudy and me and our cousin Gertie Hoffmann. Johns and Hicks hired a Chev Tourer in Adelaide and joined us for the journey. In those days you had to have a permit to cross the border to go interstate. A special permit was also needed from the owner to take a hire vehicle out of the State.

We stopped at Bordertown to get the permits for travelling into Victoria. Johns and Hicks apparently stated that they owned their vehicle instead of giving the true information that it was hired, although I was unaware of that until later. While the policeman went outside to check our registration numbers they grabbed some registration stickers for the windscreen and other papers from under the counter and slipped them in their pockets. When I queried what they wanted them for they said something to the effect, "They might come in handy some time." Of course, later on I found out that they sold the hire car in Melbourne and hired another vehicle which they took to Sydney and sold there. Unbeknownst to me, they probably had that all planned before we even left Adelaide.

Trudy, Gertie and I stayed with some friends in Melbourne. The two young women were due to start work at Immanuel College in Adelaide in the new year, so they were enjoying a holiday until then. Hicks and Johns had taken a flat when they arrived in Melbourne where I joined them for a few days, then they disappeared. It soon became obvious they had no intention of coming back. They had apparently been engaged in all sorts of shady deals about which I knew nothing until some detectives arrived at the flat and asked me if I knew them, where they were and the rest of it. By that time I was looking for them too because they'd left me 'holding the baby' for all the rent at the flat! Fortunately the police believed me when I said I knew nothing about their illegal activities.

Hicks and Johns hadn't paid any rent to the landlady and I couldn't afford to pay her either. I explained to her what the situation was and how I was also waiting for them to return. Hicks and Johns had left about 50 beer bottles on the verandah of the upstairs flat. 'Grog' wasn't supposed to be taken into the flat at all – the landlady and landlord had a very strict rule about that. I knew the Bottle-O was coming so I bundled about 20 bottles at a time into a sheet with a cord from one of the blinds and lowered it down to the Bottle-O below, which I cashed in to get about five shillings. As I was lowering the last bundle the landlord saw it going past his window and raced out to see what was in it. He rushed upstairs and literally threw me out, which ended my stay at the flat! From there I went to stay with Sam Weller, a friend of ours.

We heard later that Hicks and Johns had been arrested in Sydney. While in Melbourne they had formed a bogus company which they called 'Toledo Gold Mining Company' which apparently wasn't legally registered. They simply put a big 'ad' in the paper in prominent headlines saying that Toledo Gold Mining Company had found new prospects near Ballarat and they wanted partners and so forth, which is how they acquired quite a large sum of money before they absconded to Sydney. That was the crime for which they were arrested – defrauding public money, false advertising, stealing cars and the rest. Paul Johns was deported back to Germany and Hicks was jailed in Sydney, which was the last I heard of them. I think I was pretty lucky to escape from trouble at the time because at first the Police suspected I was involved with them. However, I had enough evidence to convince them that I was only visiting Melbourne. They took my name, address and where I was staying, and must have been satisfied since nothing more was heard of it.

McCartneys in Melbourne were struggling along in the Depression and like so many others, had no money. They couldn't afford to pay me back for the tyres, but they gave me ten shillings to tide me over, telling me to come back in about three days, which I did. I was given another ten shillings, and Mr McCartney told me about a cafe nearby where I could obtain a three-course meal for a shilling. It was mostly cabbage, potato, fish and stewed apples or pears and custard, but it was good food and plenty of it. With a pint of milk costing five pence, that was my meal for the day. I was stranded in Melbourne for about three or four weeks.

Finally McCartney told me he couldn't pay me the rest or give me any tyres because he was "closing up shop." I then met up with the Reverend 'Skipper' Partridge, a minister from the A.I.M. who asked me to help him deliver some furniture to a poverty-stricken family out in the country. I told him I could help if I had some petrol, and told him of my plight. He suggested I go to the Vacuum Oil Company and book up some petrol since I had an account with them in Alice Springs.

Next day I went to the oil company and obtained a 200 litre drum of petrol plus I filled up my tank, putting it on my account. Vacuum Oil also wanted some drums shifted to another depot in Melbourne, for which they paid me £4 in cash. I then shifted the furniture for 'Skipper'.

Once I had my petrol I loaded up and headed for home with £4 in my pocket and a passenger, Bill Stone, who wanted to go back to Alice Springs. I was still desperately in need of the two tyres which had been owed me by McCartneys, but it couldn't be helped. About 80 miles out of Melbourne, late in the evening, one of the front tyres which had worn right through had a blowout and there I was with no spare. I was just about to walk across to a nearby farmhouse when I saw the headlights of a transport approaching. In those days a 'big' transport was about a 20-foot semi. The driver pulled up and asked us what the trouble was. I explained to him how I came to be without a spare. He was a kind man and gave me a spare off the back of his truck which we put on my rim. He told me to call into Horsham where he had a depot. He said, "Go to my garage and tell the man that I've sent you and you can have the retread tyre. Swap it for the one we've just put on because I need that one. You can have the retread and fix me up some time." He wanted £8 for it.

That was the sort of kindness that I was accustomed to up north in the Bush, but I didn't expect it to happen down in Victoria. When I returned to Alice Springs later I wrote to him thanking him very much and sent him the money. Later I had a lovely reply of thanks back from him, saying he had never expected to hear from me again.

We were heading towards Adelaide via Mount Gambier. When we arrived in Kingston the police there didn't want to let us go through alone to Tailem Bend. From Kingston to Tailem Bend was what was called the 'Ninety Mile Desert' which was a sandy track through the sandhills and mallee. The police advised us there should be at least two vehicles travelling together. However, when I told them I came from the Territory, was well-equipped and used to desert driving, he let us go. When we reached Tailem Bend we notified the police that we had arrived safely, then travelled on to Adelaide.

In Adelaide we were joined by another young chap, Bob Gregory, who also wanted a lift to Alice Springs where his family lived, so I had two passengers on board. Of course, we were all out of money except for a few shillings. I asked if either of them had any food, which they didn't and neither did I! Before we left Adelaide Bob suggested we go to the East End Market on Saturday, just after midday, when the merchants wanted to clear out and would sell their produce off cheaply. He knew a chap there for whom he occasionally did odd jobs. We thought we'd be able to get some cheap vegetables such as cabbages, half-damaged potatoes, onions and bits and pieces. However, all that was left by the time we arrived were a couple of loaves of bread and two half-cases of sweet Christmas apples which became our main diet on the way home. We ate apples until we looked like them!

Next morning we set out on our return journey from Adelaide to Alice Springs with Bill riding on the back of the truck. About 50 miles out of Adelaide, just after dark, Bill's hat flew off in the wind. He yelled at me to stop so I put the brakes on. He called out, "I've got it!" so I released the brakes again. He thought I intended to stop completely, but we were still moving at about 10 miles per hour when he hopped off sideways, hit the ground and spun around with his leg going under the back wheel. Luckily his leg was flat down and the wheel went over the inner side of his leg which caused very bad bruising but his leg wasn't broken. So we had an invalid all the way up. We reported the accident at Clare hospital where the staff didn't want to let Bill travel on, but he refused to stay because we were all broke.

The borrowed retread from the chap in Victoria carried us all the way home. I purchased another drum of petrol at Quorn and sent a telegram to the Vacuum Oil Depot in Alice Springs to see if they needed any loading brought up. They said "Yes", so I loaded two tons of petrol at Quorn to help pay my way back home. I would have received £16 for that loading.

Unfortunately, halfway to Oodnadatta we struck some rain and became bogged. In getting out of the mud one tooth of the pinion in the differential was busted. With much jacking and lifting up out of the bog I was able to readjust the pinion so that it bit right in tight to the crown wheel and didn't slip when it got around to the broken part. We slowly made our way into Oodnadatta via Crows Nest, going, "Toc, toc, toc, toc!" every time the pinion went around.

When we reached Oodnadatta we had to unload the petrol destined for the Alice at the Vacuum Oil Company there, which was bad luck because that meant we didn't get any freight paid on it. We stayed at Oodnadatta until another pinion was sent up from Wagner's Wreckers in Adelaide the following week. People at Oodnadatta were very hospitable and gave us help while we stayed there. Eventually we set off again and finally arrived back in Alice Springs – the end of quite an eventful holiday to the 'Big Smoke'.

Chapter 16
Aulgana Mica Mine

When I was in Melbourne staying with Sam Weller, who owned a company named General Mica Supplies, he mentioned that he wanted to go back to the Territory and start mining the mica again. He had been up there mining mica a few years earlier and suggested that if I could build an air compressor to take up to the Aulgana Mine, he would buy a jackhammer, hoses, drills and other gear and join me later. The air compressor would need to be in a dismantled state since it would have to be carried up a walking track to Mount Palmer.

The compressor I built was made up from two old Dodge 4 motors. I cut one motor in halves and made up two special cylinder heads to fit on the half motor which became the compressor. It worked quite well though it wasn't the most efficient compressor in the world, but it was much better than using a hammer and hand drill. Up to that time all the Italians working in the mica mines were drilling by hand.

Sam and I headed out for Harts Ranges in 1937 and established a camp at the bottom of Mount Palmer. The Aulgana Mine was about 300 metres above the camp. All the petrol and machinery had to be carried up and the mica brought down on our backs. We worked with two Italians who both came from the mountain country in Italy. They knew all about carrying loads up and down mountains!

Camp at Aulgana Mica Mine

Eventually we mounted the compressor up near the mine. One part, the air cylinder, weighing 120 kilograms, was the biggest and heaviest part to be taken up. It was carried up by 'Big Peter' Pizzinato, a 6 foot 3 inch stalwart fellow.

During the eight or nine months we worked there we obtained quite a large quantity of good quality mica, with several quite large books weighing over two tons. The books had to be split, then carried down. Every morning we carried up petrol and water to the mine and mica down in the evenings. The Italians showed us how to make up special saddles, similar to the ones they used in the mountainous country in Italy. The saddles fitted onto our backs with straps over the shoulders. After all the big books were trimmed and sorted there was about a ton of marketable mica from them, worth about £2,000, which put our bank balance in the black.

Samples of large books of mica, held by Strappazzon at the Spotted Tiger Mine

About eighteen Italians had established themselves at the 'Spotted Tiger' and some other mines such as the 'Disputed', the 'Caruso' and the 'Billy Hughes' mines. Some of those were on the mountain top too. About every couple of months the Italians would get me to bring out a load of stores for them and cart their mica back to town. One of the main staples of their stores were big barrels of "grappa" or wine!

When the barrels of wine were delivered everybody would help with bottling it and they'd have a party. Naturally quite a bit of it didn't get into the bottles! The amazing thing about some of these men living in camps, often 2 or 3 miles (5 km) away in the hills, is how they'd find their way home at night with only a faint kangaroo pad to follow from one camp to another. After the party was over, perhaps at one or two o'clock in the morning, they'd head off in different directions, walking home half drunk with only a hurricane lamp to light their way.

I heard a funny story about one of them who missed the track, became lost and just went to sleep out in the hills. It was all mountainous country and when the fellow woke in the morning he wondered where he was for a while, but once he'd climbed a peak it was possible to see where the camp was and he went straight down from where he'd slept back to the mine.

The Italians were an interesting group and often talked to me about their experiences in life and where they had been before and how things were in their 'old country' Italy. Of course, one topic which was always on the agenda was a discussion about who had the best mica, the biggest books and so forth. In later years, when they acquired air compressors and jack-hammers, they often called on me to fix their mechanical problems. They would say, "German brains good for engines. Italian brains good for mica!"

As we tunneled into the mountain following a reef of mica, we had penetrated in about 200 feet when we struck a stream of water. It turned out to be a little spring which ran continuously. That saved us carrying water up every day to the mining site for drilling. Every time we went down, if we weren't carrying mica, we carried some of that beautiful, pure water down to the camp in a 4-gallon drum. That was quite an experience working out there. Up on the side of the mountain we could see probably 50 miles (80 km) or more out in three directions. The only direction we couldn't see was east because the mountain blocked the view.

Incidentally, the Aulgana mine was the last mine Bill Petrick, my brother-in-law, had worked in the 1920's before he gave mining away, married Elsa and took up a station property at Mount Swan, north of there. While he was working up there he'd watched the thunderstorms going north over that area, always coming across at the same place and decided it would be a good property to own since it would get moisture from the storms as well as the regular rainfall.

In those early days of developing those mica mines it became known that the Harts Range Mica fields were the largest in the world. Burma has some large mica mines also, and it was very hard to compete with them because of their cheap labour, but the quality of the Harts Range mica was very good. It hadn't been mined out at that time and was all close to the surface. One of the problems in prospecting for mica is that you might see some reefs in the distance, but when you'd arrive and check them out, there would only be small mica, as in black mica.

The Italians were very successful in locating the best places for mica mining 15 or 20 miles (30 km) into the mountain ranges. The first thing they did after finding good mica, was to make a road in. Half a dozen of them ganged together and worked for a couple of months building a road in to it. Before they acquired trucks they had used camels, making a camel pad into the places where they set up camp, which was much easier than trying to build a truck or jeep track into those areas. Towards the end of the time I was working up there, Dad hadn't been very well and I had to leave the Aulgana mine to take care of things back in town. Gradually the mica supplies cut out and they never got very much more after I left. They would joke about it and say that I must have put a 'hoodoo' on the mine. The price dropped and we gave the mining away, but it had all been good experience.

Chapter 17
Rough Roads

Around 1930 a chap by the name of Charlie Deakin, a road engineer from Tasmania, was employed by the government as Roadworks Supervisor. He was given the task of making a road through the Alice Springs Hills to the north. The old road used to virtually follow the Telegraph Line up through the hills on a very rough track. It could take an hour to travel 10 miles (16 km).

People today may not realize that those tracks would nowadays be classified as suitable for four-wheel drive vehicles only, which classification would have been given to most of the track for the mail run through from Alice Springs to Birdum. In the Wet, of course, it was a lot worse. The timetable for the mail in the Dry season verifies this – Alice to Tennant Creek three days and two nights; Tennant Creek to Birdum three days and two nights, a total of 600 miles (960 km). During the rainy season it could take anything from a week to a month to negotiate the track, or be impossible to travel on at all.

By 1934 a reasonable sort of track had been made through the hills by the Department of Works, some of it using prison labour. When Charlie Deakin took on contracts, he employed his own men to work on the worst parts of the road up north – places like Stirling, Barrow Creek and from Wauchope through the Devils Marbles to Bonny Well, north of Tennant Creek through Attack Creek and on to Powell Creek. The track through Powell Creek was never actually done. The road was finally diverted east of Powell Creek and was completed when World War II broke out. The old track from Tennant Creek through to Helen Springs, Powell Creek and on to Newcastle Waters was completed in its worst parts by about 1938. Although it was only about 150 miles (240 km) it would take about 12 to 15 hours to cover that stretch, if you had a good run and were carrying a light load.

I used to cart quite a lot of stores out to Hermannsburg in the 30's which had previously been carted by camel teams from Oodnadatta. In those days it was only a track of about 80 miles (129 km) from Alice Springs, but with a fully loaded 3-ton truck it could take up to ten hours to complete the trip on a good run. The old motor vehicles, having 23 or 24-inch wheels, were able to negotiate rough roads and heavy sand. The track was eventually upgraded in the early 1940's.

Charlie Deakin's road gang and equipment comprised two one and a half ton Diamond T tip trucks, and one 2-tonner, a ute, four tents, three of which were used for accommodation while the fourth one was used for storage, with a bough shed next to it as a kitchen and mess. Most tip trucks in those days had hand-operated winches to operate the tipper body. Deakin was very proud of his new two-tonner as it was equipped with a hydraulic hoist. He employed six men with picks and shovels, two men for each truck. Each truck had to be loaded by hand, then tipped out onto the road. Deakin himself would do most of the spreading of the gravel in the sandy or boggy places and at creek crossings. On this occasion, Deakin had set up camp near the Devils Marbles since there was a stretch of about 8 miles of road to be repaired there, including two creek crossings to be cemented. They were camped for about six months on that stretch. He came down to the Alice once for some equipment and supplies, but generally I carted all his perishable stores and fuel up on the mail truck.

On this visit to Alice Springs he wanted my help. His concrete mixer had broken down, with the teeth of its main drive gear completely worn right off. He asked if I could do anything with it since the Department of Works had told him he'd need to get a replacement up from south and he couldn't afford a two-week wait. This was on the afternoon before I was due to leave on the mail run. I immediately set to work on it with the oxy-welder and built up all the teeth on the gear. I was the only person in the Alice, other than the Department of Works, who had an oxy-welding plant. Deakin was very pleased with the result. He didn't even need to file it to make it fit. He just put it straight on and it worked for the next four years!

There were often 'hoboes' travelling up and down with me on the mail runs. Quite a few of them worked with Charlie for a while, but most of them didn't last very long with him because, if a man couldn't pick and shovel 4 or 5 metres of gravel onto the truck in a day within the first week he wasn't any good to Charlie. By the time a fellow had been there for three weeks Charlie would expect their average to improve to about double that quantity. Most of them only stayed about a week or a few days when they either 'tossed it in' or he kicked them out! They were paid £1 a day and given their food, which was quite good wages during the Depression.

In that period the Government bought a couple of rubber-tyred farm tractors and fire ploughs. These were dragged around the back tracks on the station roads and on the main road to help clear the grass and shrubs off the centre of the tracks and also acted as a bit of a grader. On fairly soft ground it did quite a good job. Also, around the Tablelands in the black soil areas, they filled in the potholes and the cracks. Those were the forerunners of the road graders. The tractor and the fire plough were also used to carry a bit of their gear and fuel. There was also a light tip truck running along with them as a tender for the camping gear. In places where gutters had to be filled in the two men did the job by pick and shovel and lots of sweat.

There were several of these teams of workers who travelled around the Territory, improving the roads during most of the 1930's, which greatly helped motorised travellers in more ways than one. A team of two men did duty as cooks, truck drivers, tractor drivers, mechanics and pick and shovel operators, and in their spare time they acted as laundry maids! The tracks were certainly improved and had an added benefit of saving a lot of radiators from becoming blocked up with grass seeds which could be a great menace, particularly for the uninitiated who would run their motors hot and out of water.

In about 1937 the Government bought the first caterpillar grader, which was dubbed the 'Yellow Peril'. It was a big 3-cylinder vehicle of heavy construction and helped improve the roads considerably. A chap by the name of Fitzpatrick, or 'Fitzy' as he was known, operated it for several years. He wanted to have a 'feather in his cap' for having graded the road through from Alice Springs to Birdum. He did a very good job and used the grader as a bulldozer in many places, cutting into the sides of hills and changing the road alignment. He had gone west of Helen Springs and on the eastern side of Powell Creek, up on the high ground. When the war broke out in 1939 road gangs with bulldozers and other machinery took over from him, but he had done quite a job with his mighty machine, the 'Yellow Peril'.

Fitzy always worked on his own up there. Several times he was given an off-sider but he could never get on with anyone – not even a cook or an off-sider to help shift camp. Anybody sent up would only last about a week and Fitzy would kick him out. Eventually they had to accept the fact that Fitzy was better left to work on his own. They were a bit worried about the risk if he had an accident, but he managed satisfactorily.

Nearly every mail trip I'd have some stores or spare parts on board to deliver to him. There might be half a dozen cutting edges for the grader and a tyre or a tube. A fuel truck went up about every two months to deliver fuel and water to him. Fitzy had a trailer loaded with fuel and an old-fashioned caravan. He'd hook them together to shift camp, like a little road train, and set up camp in a new location and get back to work. Passing through the different cattle stations he'd get some meat from them, or they'd run some meat out to him whenever they killed an animal. They supplied him with meat in exchange for him running the grader over a few of their worst station tracks.

One day I arrived with the mail and found Fitzy well and truly bogged with his grader. He'd struck a spring in yellow, sandy clay and the grader had sunk down about 3 feet (1 metre). He had to leave it there for about a week until the ground dried up a bit. Eventually George Nichols came along with the Government A.E.C. truck and a load of fuel for him. Between the two of them they dug it out and were able to tow it out with the big truck. Arthur Nichols, George's brother, was one of the fire plough gang who had operated the ploughs before the grader was bought. I think a couple of the other workers there were 'Taffy' Pick and Syd Branson.

Another time I was bringing the mail through just after the Wet and followed Fitzy's new track. Fitzy had filled in a deep gutter with the grader, just pushing loose dirt in. In the meantime there'd been a heavy thunderstorm which had soaked all the loose dirt. The rest of the road was

Getting out of the bog with 'hobo' help at McLaren Creek Crossing during the 1939 floods

in quite reasonable condition, but as I was coming along the front wheels just dropped right out of sight. The front of my six-wheeler Mulga Express Mark III, which was a full forward-cabin at that time, just sat flat on the ground. Luckily I had a few hoboes with me on that trip, cadging a lift and they helped me dig and jack our way out which took about half a day. Fitzy had intended coming back later when it had dried out a bit and compact it by running the grader back and forth over it but that hadn't been done yet. I backed out until I found a place where I could turn around and follow the new track back to where the old road crossed it and made my way onward from there. Those were a few of the hazards.

On the old road, after the Wet, we had to be especially careful because, with so much high grass growing in the middle of the road it was impossible to see the occasional wash-out across the track. If one didn't drive carefully the vehicle could suddenly drop into one of these wash-outs, hidden by the tall grass. We had to detour around them or set to work filling them in, chopping the banks down to get across, quite a regular occurrence after the rains.

One time around 1937 or 1938, after the Wet, a Government ministerial party was going through to Darwin and discovered such a hazard. The party included the Administrator, the Government Secretary and two other VIP's. They had picked up two brand-new Buick cars in Alice Springs and were driving them up to Darwin. About halfway, between Powell Creek and Newcastle Waters, on a plain about 5 miles (8 km) long which was one of the best stretches of road around, they speeded up. Next thing they hit one of these gutters which had washed out straight across the road. They were travelling fast enough for the first car to almost jump it, but both the front wheels and the suspension were buckled. They had cleared about a 9 foot (3 metres) wide straight up and down gap about 3 feet (1 metre) deep! It practically wrecked the car which had to be left there. When the second car arrived on the scene they managed to pull up in time when they saw what had happened. They all piled into the second car and drove on to Newcastle Waters where the Works Supervisor, Bill Carroll, organised a truck to go back and retrieve the other car. It was taken on up to Birdum, and both cars were loaded onto the train for Darwin.

Bitzer Mark III just bogged a bit on one side

Chapter 18
Building 'Bitzers'

From early childhood I had learned to improvise, repair, modify and invent solutions to mechanical problems. Dad was a good role model and I added my own particular brand of inventiveness and aptitude for such things.

The Dodge 4 wasn't large or sturdy enough for the mail run from Tennant Creek to Birdum which I took over from Sam Irvine, so early in 1936 I started building 'Bitzer' Mulga Express Mark I from bits of this and bits of that. It was completed just in time for Foy's expedition. It was a more or less conventional two-wheel drive truck built from bits of abandoned trucks. One of these was the burnt-out Willys Knight truck previously owned by Gus Brandt and Ivor Windelman.

Before I built my first 'Bitzer' I'd had plenty of experience with Dad during the 1920's and early 1930's when we'd had to improvise to keep the Old Dodge 4 on the road. Tyres were very easily damaged on the bad roads. If a tyre was staked or sustained a nasty split about four or five inches long, we'd take another old tyre, cut a sleeve out of it and taper the edges off. We'd then insert a piece of tube rubber in between to prevent chafing. Next we'd use a hot rod to burn holes through the two layers and insert about six gutter bolts on each side of the split. In most cases that would last until the tyre wore out. The last tyre repair I did using that method was when I was carting ore from Jervois to Mount Isa with the road train a couple of decades later!

'Bitzer' at the Birdum Hotel (railhead south of Darwin) on a mail run, 1937

'Bitzer' leaving Birdum with a load of crude oil for 'El Dorado' Mine at Tennant Creek, 1937

Old model Dodges, up to about 1924, had a habit of stripping the first gear because it was a bit too tight. If that happened while travelling through really rough country we could generally manage fairly well by using second and top gear. However, if there was a steep creek or hill, before going through, we'd have to turn around and reverse through.

And talking of reversing through heavy patches – the petrol tank was under the seat in T model Fords. The petrol simply flowed from there to the carburettor. There was no petrol pump, and if there was a very steep hill the petrol wouldn't flow to the carburettor. To cope with this situation, I'd let it run back, start the motor again and reverse up the grade. This brought the petrol tank up higher than the carburettor enabling the fuel to flow through freely.

While driving in the open trucks ('Bitzers' Mark I and Mark II) in rainy weather or on frosty nights, besides wearing a flying helmet, I'd take a piece of a large truck tube, cut two holes in it to see through and pull it over my head with a cut away to go over my shoulders reaching down to my chest. My goggles went over the top of that. We called them our 'Ned Kellies' because it looked similar to the bullet-proof gear worn by the famous outlaw.

Acquiring the Chassis for 'Bitzer' Mulga Express Mark I

Back in about 1925, which was about the same time as the government buildings were being started in Alice Springs, Ivor Windelman and Gus Brandt, two Swedes, arrived in the Territory with two 2-ton Willys Knight trucks. They had obtained a contract to cart building materials from Oodnadatta to Old Crown. Another chap, Harry Wolf, had the carting contract from Old Crown to Alice Springs.

About 1926 Windelman left the Territory when one of their trucks was burnt out near Horseshoe Bend. The story given to the insurance company was that when they were crossing the Finke River the truck became bogged going up the far bank. They unloaded most of their load and were ready to try and get out of the sand when a spark from the battery cable, which was touching the petrol tank, burnt a hole and ignited the petrol which in turn burnt the wooden cab and the front load of timber on the truck.

Claude Golder, who owned Horseshoe Bend Station at the time, took a team of horses down to clear the road and towed the remains back to the station for the insurance company. The insurance company weren't interested in travelling 1,000 miles (1,600 km) to inspect the truck and told Golder he could have the wreck in lieu of his towing charges of £8.

Nine years later, in 1933, I was looking for a truck chassis to build a 'Bitzer' and I remembered the burnt truck at Horseshoe Bend, which had a good strong chassis. I sent Golder a telegram saying, "If you still have burnt truck how much do you want for it loaded on rail to Rumbalara siding?" The reply read, "Can put on rail next week thirty-five pounds."

About four weeks later I was stripping it down when Ben Nicker came into the yard with his old Chevy 4 wanting its generator fixed. Ben was the eldest son of the Nicker family who had settled at Ryans Well, about eighty miles north of the Alice around 1914 – one of the early pioneer families of the Territory. Ben was a big lanky fellow, about six foot three inches, with scruffy ginger hair, who said very little until he got wound up (especially after he'd had a few drinks when he'd become the happy, wild Irishman that he was). He looked at the wreck and walked around it for a while, not saying anything until eventually he came out with a question.

"Where'd you get this heap of shit from?"

I told him it was from Claude Golder at Horseshoe Bend.

"Well, stone the bloody crows! I thought it was dead and buried a long time ago! Wasn't that old Windelman's Willys Knight?" He told me he'd been an off-sider working for Windelman when the axle had broken in the creek, which was the last straw after many other troubles he'd had under the seat with that truck. While they were unloading the timber a spark from the battery cable started a small fire under the petrol tank which had a small leak. Windelman grabbed Ben's .32 Winchester from behind the seat and fired a bullet into the petrol tank with the comment, "That will fix the f--- bitch of a thing good and proper!" Ben pointed to a hole in the tank where the bullet had entered. And sure enough, when I dismantled the differential, there was the broken axle! So that is why it was burnt.

'Bitzer' Mulga Express Mark II

The insurance company paid Brandt full insurance on the burnt out truck and Windelman took the other truck south and dissolved the partnership, as their contract had been completed. Gus Brandt stayed on in the Territory and became the main butcher in Alice Springs, with several shops. At the age of 72 he bought himself a little glider and said in his strong accent, "I vont to learn to fly! I have always vonted to learn to fly before I get too old to learn!" He joined the Alice Springs Gliding Club and learned to fly after many hair-raising lessons, but he didn't continue after he had fulfilled his life's ambition.

Around 1937 I converted my 'Bitzer' Mulga Express Mark I into a full forward control with a heavier front axle to double the carrying capacity from 4 to 8 tons. Later in 1938 I needed still greater loading capacity so I converted 'Bitzer' Mark II into 'Bitzer' Mark III by turning it into a six-wheeler with a full forward cabin and a 22-foot tray which carried up to 12 tons in weight. It was constructed out of six different makes of trucks and many other bits and pieces!

The chassis was built out of a Willys Knight and a Dodge Graham with a Thornycroft bogie, known as the brakeless model, which I modified. I doubled the springs and made some other modifications. The auxiliary gearbox was the 4-speed box out of a 3-ton Dodge-Graham which I modified. In front of that was a Studebaker gearbox with a servo-hydraulic cable brake system connected to the bogie and in front of that was a Silver anniversary Buick car engine. The front axle was a Timpkin heavy-duty axle and the radiator was from a Thornycroft. The steering box came from a Garford, one of the original old brewery trucks which were used around South Australia when horses and wagons were first abandoned, and I ordered timber and steel for building the tray and cab which was sent up from Adelaide.

Although it was a 'Bitzer' that truck did a mighty job. I must have totalled 60,000 to 80,000 miles (approximately 100,000 to 120,000 km) in it before I finally scrapped it. I had intended to keep it as a souvenir, but while I was away in later years some workmen decided to clean up the yard up in the Alice and towed it down to the tip. By the time I returned it had been completely burned.

'Bitzer' Mulga Express Mark III

Improvised Repairs

In 1939 when I was still running the Overland mail I bought an old Federal truck from Sam Irvine. It had a habit of throwing its con rod bearings, particularly on a stinking hot day, if it was over-revved a little going through a bog or a heavy patch of sand. On one trip I was near Muckaty Bore near Helen Springs when it happened again, so I pulled the sump off. There was a Comet windmill nearby so I climbed up and cut a piece off one of the blades. It was very thick galvanised iron, heavily coated with zinc, which is a slippery metal similar to white metal. I shaped a bearing out of that, fitted it into the con rod and drove on down to Alice Springs, which was about 400 miles (640 km).

If a vehicle had a broken spring, a solution was to get a piece of truck tube rubber rolled up to about a diameter of 6 inches, and place that between the chassis and the spring. The elasticity of the rubber would provide sufficient cushioning and would last for hundreds of miles.

An emergency fan belt could be made out of a 3-cord rope. The rope needs to be a bit over three times the length of the fan belt's circumference. Unravel the rope, taking one cord out of it, then make a loop and keep wrapping it round itself, turning it into a 3-cord loop of rope with only two ends. Cut the two ends off on a long angle and overlap them using rubber solution or glue if any is available, then wrap them very tightly with fine wire or string. That makes quite a good fan belt. Since panty-hose have come onto the market, several pairs of them wrapped around, twisted and knotted together, also makes quite a good fan belt which will last twenty or thirty miles in an emergency if you don't drive too fast.

To mend a badly staked radiator, caulk up the holes carefully with rag where the stake has entered, making it as tight as possible. Then if you have any pepper put in a spoonful with water, or get some grass seed like sesame seed, crushed up finely. Once it starts getting into the leakages and holes it becomes stuck and swells, thus blocking the leaks, making quite an effective temporary repair.

Crossing Wycliffe Creek in 'Bitzer' Mark III

Chapter 19
My Wettest Mail Runs

A couple of years before World War II, I acquired the full Alice Springs to Tennant Creek to Birdum mail contract after Sam Irvine had become too ill to carry on. It was a weekly run to Tennant Creek and a monthly run to Birdum. I finished off Sam's current contract of about six months and obtained a further contract for the next three years.

On my first mail run up to the Birdum railhead I decided to take the train from there and go up to Darwin to have a look around and see what it was like instead of just remaining at Birdum waiting for the return mail. The mail contractor was required to wait a week at Birdum before commencing the return journey in order to give people in Darwin time to respond to any southern mail.

The train stayed at Birdum overnight after receiving the mail delivery from the south and left the following morning for Mataranka and Katherine and on to Pine Creek where it stopped overnight again. On the return trip from Darwin it stayed at the same places. The old train had no lights and only travelled during daylight. Its best speed was about 25 miles (40 km) per hour. Seeing Darwin was quite exciting for me and quite different from the life I'd known in 'The Centre'.

Our old Dodge proved its worth delivering the mail many times during the Wet seasons up north. The following entries from the Adelaide Postmaster General's file recorded some of my journeys in 1939-1940 (written up in John Maddock's *A History of Road Trains in the N.T.*, Kangaroo Press, 1988, pages 103-4).

21.12.39	Telegram. "Wet season setting in early up here. Horses can only carry light mails. Regards. Johannsen, Mail Contractor.	
28.12.39	Johannsen used motor transport.	
5.1.40	"330 points rain recorded Powell Creek this morning. Southbound Overland mail held up here." 6th Jan. Overland mail still in bog one mile south of Powell. 8th Jan. Johannsen still at Powell Creek. 9th Jan. Left Powell Creek 1:55 p.m.	
26.1.40	Mail Contractor Johannsen sought permission to fly overland mail Newcastle to Daly and Daly to Newcastle (Creek uncrossable without soaking mails as boat not available). Johannsen to pay costs and arrange horsemen to pick up Dunmarra road mail if any. Permission granted.	
1.2.40	Overland mail arrived at Powell Creek per truck. Johannsen said he had to swim Newcastle River with the mail, as the boat was not serviceable. He swam the river, pushing the mails ahead on a campsheet.	

That swimming episode, reported on 1.2.40 caused me further trouble and led to an official reprimand. The report from the Acting Superintendent of Postal Services, directed to the Deputy Director stated:

1. "On or about 1.2.40, Mail Contractor K.G. Johannsen, of the Alice Springs/Birdum mail service, while in the course of conveying the mails, opened the Newcastle Waters-to-Alice Springs mail bag. The Mail Contractor reported his actions upon arrival at Powell Creek and has since submitted a report in which he stated that, owing to floods, and the fact that the boat at Newcastle River was in bad repair, he was obliged to swim the Newcastle River and, in order to have the whole of the mails in one parcel, so as to facilitate a safe crossing, he opened the mail bag for Alice Springs and enclosed therein all the other mail bags and floated the one bag containing all the mails over the river on a raft to which he had fitted a camp sheet.

2. The Postmaster, Dale, Alice Springs, in forwarding the Mail Contractor's report, advised that his enquiries from the Police Constable at Newcastle Waters elicit that the boat was in good order and was actually used by the Mail Contractor. Mr Johannsen, however, maintains that the boat was not serviceable and among other defects there was only one oar available and he did not use the boat but had to swim the river.

3. Paragraph 28 of the General Conditions of Mail Contracts provide that any Contractor or Mail Driver, who contrary to his duty, opens or tampers with any mail or postal article is guilty of an indictable offence and liable to a penalty not exceeding £100 or imprisonment for not exceeding two years, vide Section 115 of the Post & Telegraph Act.

4. There is no reason to suppose that the Mail Contractor opened the mail bag for any dishonest motive. His action, however, was most irregular and I am much concerned regarding the serious error committed by him. There are, however, some extenuating circumstances in that torrential rains had fallen and the country was flooded and the mail route was almost impassable, in fact, at this particular time on the Northbound trip, the Mail Contractor found it impossible to proceed further than Newcastle Waters by road and with the concurrence of this Office conveyed the mails at his expense thereon to Daly Waters and return to Newcastle Waters by air and it was immediately after this incident that he found it necessary to swim the mails across the river and committed the serious irregularity of opening the Alice Springs mail bag to enclose all the other mail bags therein.

5. In view of the peculiar set of circumstances and the trying time experienced by the Contractor and his special efforts to maintain the service, it is thought this is a case where the infliction of a fine or more serious action might be waived and it is proposed, subject to your concurrence, to specially draw the Contractor's attention to Clause 28 of the General Conditions of Contract and issue a severe reprimand to the Mail Contractor and to inform him that he must not open a mail bag in future under any condition."

Postmaster Dale was new in his job, "a new broom" so to speak, and generally followed my progress by 'phoning through to the stations en route and reporting to headquarters if I wasn't on schedule! He had a few lessons to learn in the realities of life for people living in the Territory.

An Even Wetter Run

In 1940 we had extremely heavy rains when the Wet season began and it ended up being one of the wettest seasons on record up to then. There were 317 mm recorded in January and 116 mm in February. All the rivers and creeks were in flood. I tried to convince Postmaster Dale in Alice Springs that it was no use me leaving then because we wouldn't get through, but he insisted "the mail must leave on time as scheduled." Dale was a real stickler for following all regulations according to the book, and if it wasn't done exactly right he'd invariably report it back to Adelaide. Anyhow, I left on schedule on a Sunday morning. The first 100 miles (160 km) went pretty well. The creeks were well up around 2 feet deep, but we had no problems in these because the creek beds had recently had rocks put down in them.

Once we left Ti Tree we came to Hanson Swamps near Central Mount Stuart. I skirted them without much difficulty but the Stirling Swamps were a difficult proposition. There was no way of getting around them with the water nearly a metre deep in places. After searching around for a suitable crossing, we eventually located the narrowest place which stretched about 300 metres wide and for going through the deep water I 'snorkelled' the air intake with some hose. It took us about four hours to get through, only getting bogged once.

'Bitzer' mail truck bogged alongside the historic graves at Barrow Creek Post Office.

From there on the next 20 miles (32 km) took us two days, finally arriving at Barrow Creek after many bogs. In one bog one side of the truck was right down with the tray touching the ground. My off-siders and I had to unload everything with extra help from a team of hoboes also on board. We had to jack it up, fill in the hole with rocks and timber until we finally extricated it from the bog. Then we had to reload all the mail, fuel and general stores for stations along the way. We travelled on about another 15 miles (24 km) when a similar thing happened and we had to unload and reload again. A perishable truck, driven by Murray Baldock (a 3-ton truck) caught up with me just north of Barrow Creek. From there we travelled on together and bogged down together, helping each other out. We continued to get bogged right through to Tennant Creek, taking five days of digging and jacking and unloading and reloading. With all that struggle, all the thanks we got from Mr Dale was a "Please explain" from the Postmaster-General in Adelaide as to why we were so slow getting up to Tennant Creek! Dale had reported me to Adelaide for being four days late arriving at Tennant Creek.

During the Wet season the mail truck normally terminated at Helen Springs, or Powell Creek if it was very wet. My sub-contractors, Jack Althouse and Norman Stacey, usually took the mail on by horseback. Jack took the mail from Powell Creek to Milners Swamp, and Norman Stacey took it from there to Birdum.

At Tennant Creek I received a message from Jack Althouse demanding double the normal rates to take the mail on from Powell Creek. Because of the extremely heavy rains he figured I wouldn't be able to take the mail on and he could get away with his demands. He even threatened a shoot-out if Stacey did it at the old rate. I refused to be 'stood over' and decided to complete the run myself with the old Dodge ute. I had already fitted a water-proof box on the back to carry the mail bags, my swag and rations and carried a 100 metre rope. I had also fitted 24-inch dual wheels on the back with truck chains over them to get a better grip in the mud.

I headed off with Alec Conway, my young offsider, and we travelled as far as Phillip Creek which was flowing strongly, but not too fast. We just staggered through that by putting a 'tarp' (tarpaulin) over the front of the ute and charging into the water. That was about 4 feet deep but only a narrow dip about 30 feet (10 metres) across.

The track disappearing into the flooded Lake Woods

It took two days to struggle through to Powell Creek after becoming bogged many times. After Powell Creek the road usually followed along under the telegraph line near the edge of Lake Woods. As I approached Lake Woods I was confronted with water spread out over 16 miles (25 km) wide. I had to skirt around the lake and over ridges to get back to the Overland Telegraph Line which was actually hanging in the water in some places in the flooded lake. I had to find my way up on the sides of the hills, checking every gully to find a hard enough spot to get across in my little truck.

Eventually I reached the aerodrome near Newcastle Waters after travelling along the ridges for about 30 miles (48 km), which took all day. From the highest point of the hills I could see Newcastle Waters and the little shed at the aerodrome which was just out of the water on a bit of high ground on my side of the river. However, to reach it I had to cross about 2 miles (3 km) of the overflow waters of the Newcastle River. When I finally arrived at the aerodrome fence I had to cut the wires to get onto the 'drome since the fence disappeared under the water on the lower ground. After mending the fence I made my way along the airstrip which was about 18 inches (45 cm) under water in the centre. In some parts the river had spread out over 10 miles (16 km) wide. Looking from the top of the hills it had the appearance of a massive lake.

Charlie Schultz, often called 'The Gentleman Drover', was heading back to Newcastle Waters, after droving 1,000 head of cattle to Dajarra in Queensland where he usually remained during the Wet season. He'd had a lot of trouble with his horses getting bogged on the way up from Helen Springs. When he saw my tracks he thought to himself, "Oh good! I'll follow this chap's tracks and I won't have any more trouble bogging my horses."

My bus was so web-footed with its high 24-inch dual wheels it could sneak across a lot of the softer ground just on the grass roots. At the first ti-tree gully Charlie encountered, his horses bogged down up to their bellies! Two days later when he arrived at Newcastle Waters he was amazed to find that my tracks had ended up going into the water heading towards the Newcastle Waters aerodrome, showed up again at the shed, then disappeared again into the water of the main channel of the river! Of course he didn't know what had happened, but that was where I floated the vehicle across on some drums.

I'd borrowed twelve empty 44-gallon aviation fuel drums and constructed a raft from mulga rails and wire from the aerodrome fence. I crossed Newcastle River by tying a rope to the raft with the ute on it, and then swam from treetop to treetop, pulling then securing my floating vehicle. I then swam back to undo the second rope and would swim to the next treetop. The river was about half a kilometre wide where I crossed it. Unfortunately, Alec Conway couldn't swim. He sat on the raft and gave orders declaring that he was the 'Captain of the ship'. I wasn't much of a swimmer either, but by the time we reached the other bank, having swum to and fro from treetop to treetop for about four hours, I was a much better swimmer!

While sitting on the raft about halfway across, having a lunch break (a tin of corned beef and some dry damper) word had got around the township of what I was doing. Everyone came down to the river to watch but not one of them had a camera since the films were in the mail bags I was carrying. Eventually we reached the far bank where, with the help of some of the spectators, we got the ute off the raft.

Then another problem loomed in the form of a very drunk Jack Althouse who tried to prevent me from taking the mail any further. I solved this problem by 'dropping' him with a good push. He fell over and I headed off to the Post Office amid loud cheers from the crowd and dropped the mail off. This was followed by a well-earned beer and feed at Max Schauber's Newcastle Waters Pub.

I then set off for Stuart Plain which took the overflow of the Newcastle River in high floods. At this stage it was a huge, shallow lake about 12 miles (20 kms) across. With the throttle jammed open the faithful old Dodge 4 with its improvised chains on the dual 24-inch rear wheels, slowly pushed its way across the flooded plain. The water flowing through the bottom of the radiator kept the engine from overheating as my vehicle struggled through the 18 inches (45 cm) deep water and black soil mud underneath.

Captain Swofield, a Qantas pilot who was flying an old Dragon Rapide plane on his weekly run from Daly Waters to Queensland, was a witness to this event, as he saw the wide bow wave created by the Dodge on the flooded Stuart Plain. He altered his course to investigate and radioed back to Daly Waters that the mail was on its way. Looking down from a height of 1,000 feet he'd thought it looked like a small barge or a boat. (These episodes were also recounted in John Maddock's book, *A History of Road Trains in the Northern Territory*, Kangaroo Press, 1988).

That afternoon I met Noel Healy with a plant of horses near Milners Swamp, about 8 miles (13 km) south of Dunmarra. He said, "You'll never get across the swamp. The water is up to the horses' bellies." I said, "Oh well, that's it then!" Little did Healy know what I had already come through!

Twenty-four hours later I was at Rodericks Bore, north of Daly Waters, where Norman Stacey was waiting with another lot of pack-horses to take the mail on the remaining 25 miles (40 km) to Birdum, but having come that far I decided I didn't need him to do that. I wanted to finish this epic journey myself!

Crossing McClaren Creek in the Studebaker during the 1939 floods

Chapter 20
More Colourful Characters

A couple of colourful characters that I encountered during my mail runs in the 1930's were Albert Slight and his wife. They were originally from Queensland and ran a bi-weekly transport service from Mount Isa to Tennant Creek using two Bedford trucks painted black and white, kept as neat as a new pin. They were the same themselves – always tidy and nicely-dressed – a very good-looking couple.

On a Saturday afternoon when they were in town they would get themselves all 'dolled up fit to kill' and go down to the local pub where they played billiards, boozed and danced to the party music. As the evening wore on gradually they would become a bit worse for wear from drinking. One night a woman put her arm around Albert's shoulder to which Mrs Slight took great exception! She fronted up and railed at the woman until next thing a brawl was in full swing.

On another Saturday there might be a fight between Albert and herself. She'd stand off and fight like a man. If anyone tried to come between them to calm things down a bit they'd be in trouble too! That would precipitate a double-barrelled attack on that person from both Albert and Mrs Slight. It wasn't uncommon to see them on a Sunday morning each sporting a black eye. Although they were both terribly hot-tempered they also cooled off pretty quickly.

One time, when I was staying next door to them, I overheard a 'blue' going on between them over something or other and Albert threatened to back his semi-trailer through the house. I waited with bated breath, listening for an almighty crash, but he didn't actually do it. He raced the semi backwards within a foot of the front door and stopped, went inside and started to sing, "Rose Marie I Love You!"

About two years later Mrs Slight produced a son. The constant fights changed character then, and became sparring matches about how the son was to be brought up and who was looking after him and so forth. However, about twenty years later I met up with them all up in Townsville where they had built up a large earth-moving business. The marriage had survived intact and the son had also survived it all and grown into a great burly fellow, well able to fend for himself!

Trouble saying "No!"

One day I called in at Zena and Harold Williams' store in Tennant Creek. They were already parents to three children and Mrs Williams was pregnant again. Johnny Staunton was teasing Zena about having a big family and said to her, "You must really like children, Zena."

Mrs Williams, who stuttered quite badly answered, "W-w-w-well, it's n-n-not that I l-l-like that m-m-many, b-b-but by the t-t-time I s-s-say 'N-n-n-no t-to m-m-my old m-m-m-m-man, it's t-t-too l-l-l-late and I'm p-p-pregnant again!"

* * *

"Jack the Dirty Man"

Johnny Staunton was the water carter at Tennant Creek. He had a big Dodge tanker which he used to cart water from the Old Telegraph Station for Tennant Creek township and to most of the field, barring a few odd ones who carted their own. He had been nick-named 'Jack-the-Dirty-Man' because when he worked around a truck or carried out any similar kind of work, apparently ever since he was a youngster, he simply couldn't keep clean for five minutes. He'd end up black grease and dirt all over.

He was quite an interesting character and a likeable bloke who had formerly operated solid-tyred trucks in Queensland over in the Cloncurry and Camooweal districts. When he was out on his water-carting many of the people at the different camps would have a bit of a chat with him and make a cup of tea, or invite him to join them for lunch.

I'd often help him out with a few little jobs and came to know him quite well. One day I met him when he'd come down to the pub for lunch. We ate our meal together and after he'd eaten he said, "Well, I'd better go and clean up. I've got my afternoon run to do."

I jokingly commented, "Crikey! What's wrong with you, getting all cleaned up?"

He answered me quite seriously, "Well, this afternoon's run, some of the ladies at the camps that I cart water for are a bit 'hard up', and usually one of them wants to pay me by 'favours'. I've got a couple of favourites too and I always have a bit of a clean up before I go on this trip."

The Striped Blazer Joke

Between mail trips I often took on odd jobs such as taking loads out to stations. One particular trip I went out to Coniston, north-west of Alice Springs. Just beyond Coniston near Brookes Soak a government survey camp was out there involved in mapping mining and pastoral properties.

Their chief was quite a handsome young blade and loved to wear colourful, striped blazers. Some of his gang included Bill Colson, Ben Nicker and three or four other men. The chief took a very high moral stand and thought it was absolutely shocking that these bushmen would befriend some of the aboriginal women and for their favours would give them a coat, pullover or shirt.

Because he made such a fuss about it, being a "goodie-goodie" as the men termed it, they decided to play a joke on their chief. They took one of his old blazers and gave it to one of the young women at Brookes Soak. Of course she was proudly parading around in it when he came back to camp which the men thought was a great joke, and when he came into Alice Springs they chided him about it in the pub. They knew and he knew that it was a 'put up story' but it was always presented as a great joke that his 'girl friend' was out at Brookes Soak wearing his old blazer! He had a hard time living it down.

Some people might find it hard to appreciate the fact that, for men who had lived in the bush all their lives, it was quite normal practice for them to seek out aboriginal girl friends. There were no opportunities to meet white girls because there simply weren't many available. Some of the old 'bushies' lived with an aboriginal woman all their lives and looked after them and their families. It was only after 1930-1935 that there were more young white women coming up north. There was probably a ratio of four or five men to one woman in the Outback until after World War II.

In the early days some aboriginal men esteemed it quite an honour if a white man approached him and asked for a loan of one of his wives, for which they would pay him or give him something. However, if a man didn't ask and just sneaked around and 'pinched' a woman, her husband might become quite riled about it. For an aboriginal man his women were his chattels and many of them were allowed to have more than one wife in the early days.

Returning from a mail trip with the broken-down Studebaker, 1939

Chapter 21
A Disastrous Trip

A disastrous trip occurred in 1938 on one of my runs up to Birdum with the overland mail. As the old saying went when someone had a run of bad luck, "I must have killed a Chinaman!"

A chap who was cutting ironwood sleepers about 30 miles (48 km) west of Brocks Creek, south of Darwin begged me to cart about 1,000 of them into Brocks Creek for him because he couldn't cart them in quickly enough. He was having trouble with his trucks and if he didn't deliver them within a fortnight he'd lose his contract. I agreed to do it.

After dropping the mail at the train in Birdum, I immediately set off with my three off-siders, Wauchope Tilmouth, Alec Conway and Pat O'Malley. I'd picked up Pat O'Malley in Tennant Creek and he was working for me on this trip. To reach the timber cutter's camp, we had to negotiate a terrible track, which had scarcely been used before, in sub-tropical country with very steep, bamboo-lined creeks. We finally arrived at the timber sawmill late in the evening and were taken to where the sleepers were ready, cut and stacked.

We started to work bright and early the next morning with the wood-cutters helping us to load. The contract price we'd agreed upon for carting them was 2/6d per sleeper. By about ten o'clock the sleepers were loaded and we returned to our campsite to collect our swags. There had been a very heavy dew the previous night so we'd left them open to dry out. When we reached our swags they had changed colour from khaki to yellow! They were covered with a mass of maggots – millions of them! We had to burn those swags. All we could save was our groundsheets. We hadn't been warned never to leave our swags unrolled up there. At that time of the year, when the climate was warm, humid and gooey, hundreds of dead buffalo carcases had been left lying around the area after the shooters had been through. Consequently the blowflies had proliferated and were very bad.

Leaving all that behind we set off with our load of 120 sleepers, about 12 tons. We had travelled about halfway to the Brocks Creek railway siding when we encountered a creek with a very steep bank, almost straight up and down. Nearly half the sleepers slid out from the back! It took us about a couple of hours there, carrying sleepers back up the bank, then securing them back onto the truck as best we could.

The next creek we came to also had a steep bank on both sides which threw all the weight forward onto the front wheels. It was boggy at the bottom and the front wheels pushed down into the soft mud. The wheels simply splayed outwards, buckled the steering arm and bent the front axle. It took us about half a day to straighten things up and get the truck out, leaving some of the load back in the creek. Of course, now we knew the reason the contractor didn't want to cart the sleepers himself – that track was in such a terrible state. After those trials and tribulations we eventually arrived at Brocks Creek, unloaded the remaining sleepers from our truck and decided to go on through to Darwin to properly repair the front axle and steering.

Alan Moyer up in Darwin let me use his workshop and we made up quite a strong axle using a bit of railway line. Pat O'Malley disappeared off to the pubs, became quite drunk and unknown to us, lay under my truck and fell asleep. It was the night we were due to leave and Wauchope, Alec and I had been unsuccessfully looking for him. Since we couldn't find him we decided to leave anyway so we hopped in the truck and started to drive off to go back to our camp. Unfortunately, I ran over O'Malley and killed him.

It was a terrible shock feeling something underneath the back wheels of the six-wheeler, getting out to have a look and seeing him squashed. As I'd started pulling away, just before I realized I'd run over him, a police car had come up just behind us. The road was too narrow for them to pass so their headlights were shining underneath and they saw what happened, which probably saved me a lot of complications. When he was identified to them they told me there was a warrant out for his arrest. Apparently he was wanted in Queensland on suspicion of armed robbery. They commented, "We wanted him alive! But anyhow, we've got him now, so we can scrub that warrant."

It took me a few days to recover from the shock – it was such a terrible feeling. I was only 23 years old and had never been closely involved in any death before. The only dead person I'd ever seen before was an elderly gentleman who'd died from a heart attack out near Palm Valley on one of Bond's tours. On that occasion all the formalities had been taken care of by other people.

O'Malley's death, along with the previous hassles, contributed to our deciding to give up carting the sleepers. After the inquest we set off back over the rough track to Katherine and camped by the river before we headed off early the next morning for Birdum where we loaded on 10 tons of crude oil for the 'El Dorado' mine at Tennant Creek. About 10 miles (16 km) past Daly Waters we jolted into a sharp gutter which had washed across the track but was hidden by the tall grass. The new axle we'd just installed snapped in halves. The front of the truck suddenly dropped and skidded along the ground with the two front wheels splayed out sideways. All the oil drums shot forward, bending the cabin right in! Only for two stay-bars held precariously in place by 3/8 bolts, we would probably have been sheared in halves by that load. There must have been a 'Jonah' around on that trip.

We left the truck where it was and Wauchope, Alec and I walked back to Daly Waters. I suggested we fly back to Tennant Creek by Guinea Airways and pick up the other mail truck which Dad was using to bring the weekly mail up to Tennant Creek from Alice Springs. The other two chaps backed out saying, "No way! We've had enough bad luck on this trip. The plane will probably crash if we get on it!" They decided to wait for a lift down with someone. I boarded the plane and flew home where I made up another axle and took it up with me to repair the six-wheeler on the next mail trip in my old Federal truck.

On the trip back to Tennant Creek with the mail I called in to give my condolences to O'Malley's wife and explain to her what had happened. I needn't have bothered really, because all she said was, "Thank goodness he's gone! That'll save me keeping him in booze and cigarettes for the next thirty years!" However, that didn't make me feel any better about the circumstances of his death.

When I finally delivered the oil to Tennant Creek it was the end of a disastrous trip.

Actually, the following trip when I brought the mail down wasn't all clear sailing either. I was carting another load of oil for Tennant Creek. I'd left Birdum and had travelled about 30 miles (48 km) as far as Rodericks Swamp which had quite a few boggy patches. I became caught in one place and the spine shaft in the rear gearbox twisted off.

Right alongside the bogged vehicle were a couple of hollow trees containing some wild honey-bee hives. These bees don't sting but they crawl! This mishap occurred in stinking hot, sultry weather and the bees were crawling everywhere – in my eyes, ears, down the neck and everywhere else while I was trying to work underneath the truck. They were probably after the salt and moisture from my perspiration.

I had sent Alec Conway on ahead, walking towards Daly Waters to ask Bill Pearce to come out and pick up me and the mail. We left my truck there and took the spine shaft into Daly Waters to repair it. I then borrowed Bill's ute to take the mail down to Tennant Creek where I again had to use the other mail truck. Of course all this meant that the mail was delayed in reaching Tennant Creek by about 4 hours, even after travelling all night. Naturally, Postmaster Dale was on the job, checked on all that and reported it to Adelaide.

I welded up the broken shaft and returned to Rodericks Swamp to install it the following week. By that time the bog had dried up but the bees were still around. In the meantime there'd been rain further south and Dad had become bogged about 20 miles (32 km) from Tennant Creek. I went down to pull him out and found another truck also bogged several hundred metres ahead of Dad. I walked up there to see if they had any bog mats we could use which they didn't, but I noticed a Trewhella Kangaroo Jack on the back of this truck. I immediately recognized it as one from a consignment that I'd been bringing up for the Government on my previous trip. It had slid off the load and the chap driving a truck behind me claimed he hadn't seen it when I stopped and and asked him further along the track. It was now obvious that he *had* seen it because that was the missing jack and it was on his truck! He'd painted it over in red, although it was brand-new, but underneath was the original black with the Government number on it. I retrieved it, which saved me from having to pay for losing it.

Chapter 22
Prospecting and Mining

In the late 1930's before the War, in between mail runs, I had spent a couple of weeks prospecting for rubies through the Harts Ranges. An aboriginal boy had brought in some samples of garnets and amongst them was a pink ruby. It wasn't top quality but it showed that there were some rubies in the area. The lad showed me where he had picked it up and we found a couple more little chips at the bottom of the creek sand. Because ruby is heavy it sinks to the bottom. We followed the creek up for about 10 miles (16 km) and sampled it at intervals but never found the source of the rubies.

Much later on, in about 1980, another chap eventually found the source of the rubies. They were not particularly high quality stones but were translucent with a lovely lustre. The slight impurity in them made them translucent rather than transparent. A company was formed and tons of the ruby-bearing rock was mined out and shipped away to Sydney where rubies were cut from the stone. The company went broke because it wasn't producing enough high quality stone and too much money was spent on 'chiefs and champagne parties' and not enough on the 'indians.'

On another prospecting trip before the war I'd gone out to the Strangways Ranges prospecting for virtually anything I might find. There were tales of wolfram which had been found up there by an aborigine but the area only disclosed that there was magnetite which looked very much like wolfram. After setting up camp at the bottom of the Strangways Ranges we found a few traces of lead and zinc, but nothing good enough to work on.

Then I stumbled across some strange-looking mica. I didn't quite know what it was but I thought it might be phlogopite, which is a very flexible mica. I had heard of phlogopite, which is also used in x-ray machines as a polarizer because only certain rays go through it. I did a bit of digging and found there was a reasonable-sized deposit there and later I ascertained that the market price was quite good. This was an essential material used for the manufacture of aircraft spark plugs. Nowadays it isn't used so much because modern ceramics are superior and have replaced the phlogopite.

Once we ascertained there was a market for it we set to and built a track up there. It took about a week to built the very steep track. Rather than making big cuttings we just picked a place the old Dodge could get up and it was from there that my old Dodge was named 'Euro' after the hill kangaroo. From on top of the hill it was possible to look right across about 35 miles (56 km) towards Alice Springs, as the crow flies, and on certain days we could actually see Heavitree Gap in the MacDonnell Ranges.

We carted water from Pinnacle Well and put two 44-gallon drums on the back of the Dodge which would just barely make it up the steep grades. These hills were so steep that on the way down the first time I almost panicked and jumped out of the truck because it appeared that the brakes weren't holding. Those old buses only had rear wheel brakes. It was so steep going down that the weight was coming off the back wheels! With the wheels skidding the truck gained speed, but I managed to safely negotiate around the first bend about a quarter of the way down the hill. After that it was less steep and I managed to slow down. On our next trip coming back down the hill we put half a dozen large rocks on the back of the tray to keep some weight on the back wheels, which solved the problem from then on.

We had barely established the mine and 'got going' nicely when the Allied Works Council commandeered it for the War and put thirty men on the job to mine it. The resulting output was the same as with most of those Government projects. We had mined more phlogopite in the previous three months with only four or five of us working than the thirty men did in the next six months. A helluva lot of money was spent sinking shafts and installing big machinery. We worked out that each pound of phlogopite produced cost the Government over £15.

It was then we moved up to Tennant Creek where I set up a garage before going to Brisbane.

PART III
THE WAR YEARS (1939–1945)

CHAPTER 23
MARRIAGE, WAR AND THE WOOD-GAS PRODUCER

In 1937, when I was about twenty-two years old, I met Kathleen Rowell, my first wife. She was working for Jimmy Lackman of Lackman Agencies as a cook and housekeeper. Jimmy had become quite fond of this shy young woman. He had come up North suffering from tuberculosis and expected to die up there. One lung had already gone and the other was also infected, but the climate in the Alice cured him. He played a steel guitar in bands for the dances up there. He was in his late forties and Kath was about twenty at the time. They had talked about marriage but she hadn't quite made up her mind about it.

We met when she was working at a second job in Daisy Underdown's milkbar at the Capitol Picture Theatre which had been built by the Underdowns and was leased by Snow Kenna until he built his own Pioneer Walk-in Theatre in 1942. When I asked for some iceblocks Kath warned me not to buy any when Mrs Underdown was serving. Kath said that when things were a bit slack she'd been instructed to pick up the bottles and empty glasses from the tables and empty the leftovers into the bucket, strain the flies out through a colander and pour it into the moulds to make the iceblocks! She said, "You'd better only buy iceblocks when I'm here. I always make them out of fresh lolly-water, but when the 'old girl' is here we have to make the iceblocks out of the dregs." Anyway, we became friendly and, two or three months after we first met, Kath and I started going out together.

Lycurgius (Ly) Underdown had his cool drink factory round the back of the old hotel, but it wasn't only a cool drink factory. He also made home brew. Noornee, an aboriginal, was the brewer and a funny little story comes out of that. Punch and Judy Hall recounted the tale that, when they were kids, they were playing back around the sheds. Ly Underdown hunted them out saying, "See those drums over there – those 44-gallon drums? Well, they're full of 'kadaitjas.'" The pair of them scurried off because they were frightened of 'kadaitjas' or the legendary spirits of the aboriginal people.

Well, a truer word was never spoken if you knew what was in those drums! During World War II any kind of grog was hard to get and Ly had quite a recipe consisting of O.P. rum, methylated spirits, tobacco, molasses and all sorts of other ingredients which were brewed up. So all sorts of 'kadaitjas' undoubtedly *did* come out of every bottle! Aladdin's lamp had nothing on that brew!

I was still running the Overland mail and often obtained repair work brought down from Tennant Creek. While I was repairing Johnny Staunton's Buick Kath begged me to teach her how to drive, using Johnny's car. She was getting on fairly well until I said, "Pull up here and let me drive it into the driveway." She said, "Oh, no, let me do it! I can drive in." She drove in all right – straight over Jimmy Lackman's fence post and into some wire netting! Luckily it didn't do too much damage although the mud-guard and some other parts which I'd just repaired for Staunton were buggered up again so I had to repair it for the second time!

Whenever we went out we were chaperoned by either Mum, Dad or Trudy or whoever was around, but as the old saying goes, "Love will find a way" and by the time we were married about twelve months later early in 1938 we knew all about the birds and the bees. Our marriage went along quite well at first and our first son, Lindsay, was born 18th July, 1939. Kath occasionally travelled up and down with me on the mail run until she was about six or seven months pregnant. About six weeks after Lindsay's birth he came with us too on a few trips until my contract for the mails ran out. When our mail contracted ended on 29th June, 1940 we didn't apply for it again and it was taken over by Mr L.M. Owen.

We then went to Tennant Creek where I opened a garage. It was there the idea germinated for making a wood-gas producer which would convert wood into gas to run the motor instead of using petrol. Australia was well into the War by then and petrol rationing was in full effect. I had seen an arrangement at the Central Milling Company in Tennant Creek which used wood for running a slow-speed crushing plant engine. This arrangement had one cylinder, 1 metre in diameter by 3 metres high, which was used as the burner. Another cylinder the same diameter was used as a scrubber. However, it didn't use the tar out of the wood, which gives off most of the gas. Tar was removed out of the smoke by a scrubber filled with granulated charcoal, with water percolating through it. The smoke or gas was passed upwards through the scrubber and then piped to the engine. However, it was far too bulky and heavy to use as a mobile unit of about the same horse-power.

I had an idea that if I could pass the gas from the wood through an extremely high temperature zone I could 'crack' the tar and convert it into clear gas. That idea grew in my mind as I experimented on how to make useable gas by putting wood in a retort. By doing that I produced tar-rich smoke which was unuseable in that state. I found that by running it through a white-hot tube, I could crack the tar and convert it into clear gas.

From that concept I went to the present idea, which I am still using, by creating a furnace and using some of the gas from the wood to produce a white heat which cracks the tar from the balance of the smoky gas produced from the wood. That smoky gas has to pass through the incandescent heat of the furnace and at the same time converts the charcoal produced from the wood into carbon-monoxide. This process also produces some hydrogen out of the moisture of the tar residues. Much to my surprise and pleasure, this process worked without any further alteration.

The wood-gas producer was more efficient and cleaner than the charcoal burners with which many vehicles were fitted during the War. My unit was only about half the weight and produced more power. It consumed about one and a half pounds of wood per mile (or approximately 1 kg per kilometre on a vehicle weighing about 3 tons gross) and could operate on just about any good firewood, easily obtained with a small chain saw from dead timber lying alongside the country roads. The unit I built in Tennant Creek was fitted onto a 1928 President 8 Studebaker, which I ran for about ten years until somebody backed into it, by which time it had travelled about 60,000 kilometres.

The first wood-gas producer fitted onto my Studebaker

One Saturday morning, not long after I'd finished building my unit, I was in front of Scott's Hotel. It created quite a bit of interest from the local prospectors and miners. Al McDonald, the local 'pug' and 'know-all', wanted to know what kind of wood could be used. I told him, "Anything that burns – any good firewood."

He was a bit full at the time and asked, "Do you mean you could smash up a beer case and run it on that?"

I answered, "Yes! I think it would probably do about 8 miles (13 km) on one beer case." Beer used to come in what were called "5-dozen boxes".

He said, "I'll bet you £5 you can't go from here to the Telegraph Station." It was about 8 miles away.

I said, "Oh, yes, I think I can do that."

He persisted, "Do you mean to say that one beer case would take you up there?" I replied, "Yes, and I think it might even get me part of the way back!"

An empty case was promptly produced from the back of the pub, chopped up and put into the hopper. Al and several others, all fairly well-primed and each holding a cold beer bottle in hand, hopped on board the ute and we set off for the Telegraph Station with about half a dozen cars following.

We had only gone about half a mile when Al yelled out, "Stop! Stop, you bastard! You can't trick me. You're running on petrol!"

I stopped the car, handed him a spanner and said, "You disconnect the petrol then and we'll see if I'm cheating!"

At any rate, my wood-gas producer did reach the Telegraph Station. We all pulled up there and everyone had a swig. We then set off back and were about halfway back before I had to put some more wood in. Actually, it turned out that it had used approximately a pound and a half of wood per mile while travelling at its top speed of about 65 miles (105 km) per hour.

I experimented some more and made a few adjustments on it before I decided to close down my garage at Tennant Creek and head for Sydney, via Brisbane, to put it on the market. Old Vic Foy had said earlier if I wanted to put any invention on the market he'd help me finance it. I was prevented from completing that mission and only got as far as Brisbane, but later, in 1943, I did get the Melbourne firm, Malcolm Moores interested in my invention. However, by the time they had tested it and got it 'off the ground' the War was turning the corner and petrol was priced back at 2/6d a gallon and no rationing, so nobody was interested in developing my idea.

Chapter 24

'Man-powered' in Brisbane

Tennant Creek was gradually closing down because of the war. Quite a few people from there owed me money but couldn't pay me before we left there in 1941 – they promised to send it later on. Hadley and Gitta Gates and Maisie Tindall joined Kath, Lindsay and me for the trip to Brisbane, as they were also leaving Tennant Creek plus Lassie, Hadley and Gitta's black kelpie, a dog with personality and a sense of humour.

We all piled into the Studebaker using the gas producer and headed along the old dirt track towards Mount Isa. Everything went pretty well until we were about 15 miles (24 km) from Mount Isa. Just about sundown we went into 'one of those ditches' and up the other side when suddenly the motor revved up and the Studebaker just free-wheeled.

I checked things out and discovered we had a broken axle so we pulled the cover off the 'diff' and used some of the bolts from the differential cover, plus some out of my tool box to jam the sun-wheels in the diff so they wouldn't spin around, wrapped a bit of canvas around them to prevent them falling out and continued on to Mount Isa on one wheel drive.

Luckily we were over the worst part of the road. We arrived just as it was getting dark. Next day I 'foxed around' Mount Isa for another axle with no luck. However, I found Alan Moyer, formerly from Darwin, who owned the garage. He had an oxy-welding plant and said, "Well, there it is. If you think you can weld an axle, good luck to you! You can have a go!"

I welded up the broken part and after closely examining the rest of the axle I discovered there were three more cracks about halfway through. I cut them out and welded those up too, which carried us through to Brisbane without any further problems. (Later, in Brisbane, I managed to find another axle and carried it around as a spare).

After leaving Mount Isa we travelled through Duchess to Cloncurry, down through Winton near to Barcaldine and then headed for Blackall. Between Winton and Longreach we were travelling through some open plains which were devoid of wood. I found an old railway sleeper which we chopped up and used that to fuel the Studebaker for about 120 miles (193 km) until we were back in timbered country. All of Western Queensland was in the grip of a severe drought with a lot of sheep dying. It was very hot and humid, with a few thunderstorms around.

The first creek we came to, while crossing over the bridge, we saw a lovely big waterhole underneath it. Although the water was very muddy we all stripped off and hopped in to cool off in the shade of the bridge. We no sooner got into the water when we noticed a really strong odour of dead animals but didn't take too much notice since there were a lot of dead sheep lying around the countryside. Then we felt something soft and spongy at the bottom of the waterhole. I reached down into the water to see what it was and came up with a handful of wool from a dead sheep. Apparently the entire waterhole was full of dead sheep – the water was absolutely putrid. We quickly got out, put our water drum on top of the car's roof and used a siphon hose and hosed each other down, which removed some of the stink!

We went on another 50 miles (80 km) and came to the Nive River with crystal clear water in it. We didn't waste any time getting into that to clean the smell off. We then relaxed in the water in a nice shady spot along the sandy bank. Kath noticed something was biting her leg. When she lifted it out of the water her leg was draped with about twenty big black leeches. She screamed and when we saw her leg we all hopped out of the water and discovered that we too were draped with leeches, which brought an end to our nice, relaxing swim. After those two episodes we lost all inhibitions about dressing and undressing in front of one another! We drove on a few miles and found a camping spot in a native pine forest, away from mosquitoes.

Lindsay and Lassie on the Nive River bridge, en route to Brisbane, 1941

The following day we arrived in Roma for a late lunch and parked alongside the hotel on an almost deserted street. We all trooped into the pub, tired and hungry, looking for a meal but were told we were too late because it was half-past one and meals were only served from twelve to one o'clock.

Disappointed, we all returned to the Studebaker. When I opened the lid of the wood-burner to stoke it up the local cop came over and objected to the smoke that was coming out. He said I was "making a public nuisance," and to make matters worse I had parked the vehicle nose first to the kerb instead of backing in which was the rule for all vehicles there! He told us to "Get the hell out of it before I arrest you." I think he must have been having a bad day! Or perhaps he was looking for some work to do such as arresting people or fining us! We drove on a few miles further and boiled the billy at the side of the road.

After reaching Brisbane, I ran short of money. The people from Tennant Creek who owed me money and had promised they'd send it across to me didn't, so I looked around for a job. Mars Machine Tools were advertising for fitters and turners. Well, I wasn't a fitter and turner by trade but I told them I had my own lathe back in the Territory so they said they'd "give me a start." I told them I only needed temporary work for about six weeks to save enough money to go to Sydney and put the wood-gas producer on the market. Old Mr Rapson (the business was known as Rapson and Dutton's before it was changed to Mars Machine Tools) and Vic Marsh, the foreman, said, "Oh, only too glad to help a young fella."

During my first week there I was tried out on four or five different machines – a shaping machine and lathe and finally I was put onto one of the big lathes. Little did I realize that they were only trying me out to see what I could do. However, that suited me fine because I gained valuable experience. At the end of six weeks, when I was ready to leave, I went to them on pay-day and said, "Thanks very much! I'll be off next week – the six weeks will be up. I'm giving you a week's notice."

They answered, "Sorry, we've 'man-powered' you. We've registered you as working in essential services and we can't release you."

Well, that was a blow, but I was still very happy working there and gaining good experience. I decided to pull the gas producer to pieces and modify it. Although it worked well it was originally only built up out of tar drums. In between work, it took me about two or three months to make a modified version which was more sophisticated. In the meantime I also earned extra money by going out into the country and carting a ton and a half of wood back to town with the ute. Since I had to use petrol while I was modifying the gas producer and only received a ration of four gallons a month, I decided to pull four pistons out of the eight-cylinder motor in the Straight-8 Studebaker, blocked off the oilways in the crankshaft and pulled the cam followers out. It operated quite well on just four cylinders because it was over-powered with eight cylinders. Instead of only getting 10 miles (16 km) to the gallon I was getting about 18 miles (27 km) to the gallon, which was quite a saving when the monthly ration was only four gallons!

After working about six months at Mars Machine Tools I was promoted to night shift boss which was mostly worked by the young lads. The Company was having trouble getting enough work out of the lads so they asked me if I'd take on that shift. Since I got on pretty well with the other workers I agreed, "I'll do that on condition I can still keep working on the big corner lathe." I enjoyed the interesting work coming in for that particular lathe from ships, submarines and aeroplanes. We were making all sorts of parts and part of my work was to tackle the jobs which came in with no drawings, plans or set instructions. When a job like that came in I was asked to work something out and make up a bit of a sketch of what was to be done.

My promotion and the agreement that I could continue working caused a squabble with the Unions because the rules didn't allow for a shift boss to continue working on the equipment. They wouldn't allow me to join the Fitter and Turners' Union anyway because I hadn't served an apprenticeship. The Union wanted to register me as a 'Dilutee' which meant that after the War I would no longer be registered as a Fitter and Turner and would cease to be a Union member.

I insisted that I wanted to work – I didn't want to just stand around all night. The friendly old shop steward advised, "Don't you let them register you as a Dilutee. You are one of the best tradesmen we have here." At a lunchtime meeting and vote by secret ballot with all the Union blokes, only one vote was registered against me, apparently from the Union representative who'd started all the fuss.

Everything was smoothed over and I organised the young fellows and worked out how much work they could produce, which was roughly the same amount that the day shift turned out and said, "After you have done this number...."

At first they protested, "Oh no, we can't possibly do that number."

I reiterated, "Well, after you've done that number of parts you can do whatever you want, your own jobs, or have a sleep, or whatever."

They reckoned that was great! So I was popular with the bosses because I was getting some good work out of them and I was popular with the blokes because they might only work for five or six hours but then they could do other things because they'd finished their quota.

Leaving Brisbane

In the meantime the U.S. Air Force also tried to get me released from working with Mars Machine Tools, as I was doing a lot of work for them. The pay would have been double, but we had no success. The Allied Works Council in Alice Springs also tried to get me released to install and maintain machinery in the Harts Range Mica fields, also with no success. By this time I was getting fed up. There were a lot of soldiers on leave in Brisbane, some of whom became too friendly with Kath and that is where our marriage began to deteriorate.

After about fourteen months working in Brisbane I had finished modifying the wood-gas producer and wanted to try and leave Brisbane. At three o'clock one Friday afternoon, when I went to collect my pay, I told the office workers of Mars Machine Tools that I was taking the wood-gas producer down to Sydney to put it on the market. I added, "Don't worry if I'm not back by Tuesday night just in case I get held up over there." They said that would be all right so long as I didn't miss more than a day.

By the time they realized that I wasn't returning I was back in Alice Springs. I'd sent Kath home by plane as she was pregnant with David by then. The last week I was in Brisbane I camped outside the factory in the back of the ute as I had already vacated the house that we'd rented. After Kath left, Lindsay, who was three years old then, stayed with the woman next door to our rented house until I was ready to go.

I collected my pay, picked up Lindsay and we left Brisbane that afternoon. Lindsay turned out to be a really good little traveller and mate on that trip. We travelled day and night until we arrived in Cloncurry about 3:30 p.m. the next day and went into a shop there, tired and hungry. We walked across the street to buy a sandwich and a few bits and pieces for the road. Lindsay, who was bare-footed, was already tired and grizzly and just as we came out of the shop he stepped on a soggy, red hot cigarette which stuck to his foot and gave him a nasty burn, which was the last straw for him.

After bandaging his foot and pacifying him we set off again that afternoon heading for Mount Isa, which was about 120 miles further (roughly 60 miles or 97 km to Duchess and another 60 miles to Mount Isa). I always wondered why they hadn't made a road straight through. Of course, in those days they didn't really build roads. People working or travelling just picked a track in the general direction where they wanted to go. In this case the track led to Duchess which was a very prosperous mining town, long before Mount Isa was found.

We crossed the river and went along a couple of miles towards Duchess where the track swung off to the right, going due west towards Mount Isa. I thought, "Oh good, this looks like the main track." There'd been quite a bit of work done on it and it looked as if most of the traffic had gone that way, so I followed it west thinking it would shorten our journey by about 60 miles (97 km).

We travelled along this fairly rough old track, which looked better than the Duchess road, until about one o'clock in the morning when we arrived at a mining camp, but found nobody was there. I looked around and discovered the track only continued a little way past the camp up to a copper mine. It was a dead end! Although it was only about 25 or 30 miles (48 km) away from Mount Isa as the crow flies, roughly where Mary Kathleen is now, I realised with frustration that we had to track all the way back to Cloncurry. We arrived back to where I had first left the Duchess road just before sunrise. That afternoon we finally reached Mount Isa where we had a shower, a feed and cleaned up before setting off again, camping that night near Avon Downs.

Early the next morning we set off for Tennant Creek and were within about 30 miles (48 km) from there, on top of a rise and starting to go down over the other side, when the axle that I'd welded on the way over to Brisbane 15 months earlier, finally gave way. Since I hadn't cleaned out the ash trap of the gas producer on this trip I let it cool off so I could do that after replacing the axle. I pulled the broken axle out, put the other one in and had a feed while Lindsay slept on the front seat.

There was no available wood at that spot to refuel but way in the distance, down at the bottom of this rise I could see a bit of dry honeysuckle. I decided to let the car roll down the slope for about half a mile to where the wood was. Somewhere along the way, during the night, I had lost my axe and all I had was a five-pound hammer. I'd been breaking up the snappy gum which was quite easy to break but the honeysuckle was very springy and wouldn't easily break with pounding.

I put some across the deep wheel tracks and really laid into the last solid piece. I bashed away and on a final big swipe the hammer slipped off the bendy wood which flew back up and hit me in the eye. Inside about five minutes my right eye was completely closed up and starting to turn a nice bluey-black! It was the first 'shiner' I had ever had and if it hadn't been so painful I would have been interested to watch it close up and change colours! Anyhow, I stoked up the bus with the wood I'd gathered and drove on about another 3 miles (5 km) where I found some more snappy gum. I was okay for fuel then.

Instead of getting to Tennant Creek for lunch we arrived later than I'd expected, at about two o'clock in the afternoon. I walked into Cecil Armstrong's Cafe and he greeted me with, "Hello.

How's the other fella?" making a joke about my black eye. I told him what had happened and he joked, "Oh yeah, tell us another one!" He asked, "Anyway, what can I do for you?"

"Well, I'm hungry – we're both very hungry. I'd like a good feed of steak or something, whatever you've got."

He said, "What the hell do you think this is? You're not going to get a bloody feed now. It's after two o'clock!" Apparently he was only joking, because when I saw him a couple of months later he asked, "What the hell happened to you? You just walked out on me! I was only joking!"

At any rate I wasn't in a mood for jokes at the time and simply left. We drove on another 80 miles (129 km) down to Wauchope and had tea there! Then I drove straight on through to Alice Springs, arriving about midnight. It had taken us about three and a half days to get from Brisbane back to Alice Springs, 2,000 miles (3,000 km) using about one and a quarter tons of wood and about 2 gallons of petrol.

In 1942 in the Territory, martial law was in effect. When Mars Machines in Brisbane found out that I was missing they tried to get me back, but in the meantime the Allied Works Council, which had been trying for the past six months to get me to go out and install machinery at Harts Range, grabbed me. Once I was there in Alice Springs there was no argument about it. They just said, "You've got a bloody job to do here." I worked out at Harts Range for about six months and was eventually able to convince the 'powers-that-be' that I would be of more use to the war effort if I could mine a lease I had at Strangways Range where I'd found some phlogopite around 1937.

David William Johannsen was born on 6th April, 1943 and a couple of years later, when we were back in the Alice, Peter Kurt Johannsen was born on 3rd May, 1945.

When we first returned from Brisbane, Dad, Mother, Mona and Myrtle were already out at the Strangways Range cutting and sorting mica so Kath and I and the two children moved into one of the Allied Works Council huts on top of the Strangways Range. Hadley and Gitta Gates also joined us for a while out there, prospecting.

Chapter 25
Mica Mines Revisited

I worked for the Allied Works Council (A.W.C.) at Harts Range for several months. The Government supplied air compressors and jeeps to the Italians out there to increase production of mica. The Italians built narrow little tracks right up the mountains. Instead of carrying everything up on their backs as they had done earlier they now had a jeep track about 6 to 8 feet (2 to 3 metres) wide, precariously cut into the side of the mountain. It is a wonder none of them slid over the side of the mountain or had serious accidents up there.

The remains of some of those tracks still exist. One, which was dubbed the 'Burma Road' was built by a gang of A.W.C. workers. Bob DeKane was the manager of the A.W.C. camp at Harts Range which had been established to produce mica and assist the Italians in their production. However, the only achievement Bob could claim credit for was building a road up Mount Palmer to nowhere and producing little or no mica with a gang of about thirty men.

Building that road to the top of Mount Palmer originated from a challenge which had been made alleging that "it couldn't be done." The track was wide enough to take a 3-ton truck up but it was a really 'hairy' drive. Actually it was almost useless where they put it because it didn't lead to any particular mine – it only ended up on top of Mount Palmer providing a beautiful view!

In 1944 when the Allied Works Council pulled out of the phlogopite mine at Strangways at first we were really happy that we were again allowed to produce there. The product from the Allied Works Council had been selling at ten shillings a pound in the rough. We thought we would make a small fortune at that price. We were told there would be a 'special' price for us and we thought it might be even better than ten shillings, but it turned out that they were only going to pay us 2/6d a pound after it was trimmed and cut and all cleaned up! So much for a 'special' price, although I suppose you could say it was pretty 'special' – in their favour!

Around that time a 'funny' thing happened. Actually, it didn't seem very funny at the time! At eight o'clock one morning three Army vehicles pulled up at our camp up on the hill. Three Army police officers and three privates came along, marched us all into one room and told us to sit down and wait. The others went out, leaving one officer to guard us. We asked, "What is this all about?" We were told there had been a report that we were using government-branded equipment there with brands on such items as stores, drums, tyres and the pre-fab building in which we were sitting. I found out later that there was also another vehicle down the bottom of the hill at Dad's camp doing the same thing to him.

What had sparked all this off was, on the previous weekend a couple of young officers from the Army had come out to visit the girls – Molly Reilly and my two sisters, Mona and Myrtle – and had noticed the government-branded gear. I tried to explain to them that it had been an Allied Works Council camp and that we were buying the hut and that the tyres on the truck had been put on by the A.W.C. when they had originally 'commandeered' it from us and used it while they were working the mine. But these officers wouldn't listen. They removed three tyres from the truck, several drums, a whole ute load of meat, vegetables and fruit tins, plus a half a dozen old tyres and sundry other items, as much as they could stack on their utes, and instructed us to report at the Courthouse in Alice Springs on Monday morning at ten o'clock.

When they got back to town they checked with the A.W.C. officer-in-charge, Bob Larkin, to see if our claims were legitimate, which of course they were. The tyres and some tinned food were returned to us, but they refused to put the tyres back on the truck, which would have entailed a 50 mile (80 km) trip each way plus some very red faces. Their boss wasn't very happy with them either for dragging us into town for nothing.

Dad stayed on at his camp at the bottom of the hill for another six months, cutting and trimming the phlogopite already mined out, waiting to hear the sale price. It was quite a pleasant camp and it was something Dad was able to do. Mum, Myrtle and Mona also worked there for a while. They'd all sit in the shed, cutting and trimming the phlogopite. When the low price (the 'special' price already mentioned) for the phlogopite was finally given to us, it was unprofitable to continue and we returned to the Alice.

The reason Dad and Mum were out there at Strangways Range in the first place caused them and all of our family distress and heartbreak. While Kath and I were in Brisbane during the War, a rumour had started in Alice Springs that, because our family spoke 'Deutsch' at home, Dad must be a Nazi. One evening, while Dad and Mum were attending the outdoor picture theatre some Provos from the Army arrived and my loyal, hard-working parents, who had a son (Randle) and later a daughter (Myrtle) serving in the Royal Australian Air Force, suffered the indignity of being arrested in public on suspicion of being spies. They were taken back to our home in Todd Street and had to stand by and watch while everything was turned topsy-turvy in a search for evidence of Nazi affiliation. A copy of Hitler's book, *Mein Kampf* (My Struggle) was produced and some copies of a German youth magazine which had been sent to Trudy by a German admirer. The book which had been bought at the bookshop next-door was probably being read in dozens of other homes in Alice Springs at the time!

The Army had run out of room in Adelaide House (A.I.M.) for housing the nurses and it was decided to commandeer Mum and Dad's lovely home as additional quarters for the nursing staff. Mum and Dad were told they could live out at Strangways Ranges and mine the mica. It must have been heart-breaking for them to leave the home they had built and beautified and return to a rough bough shelter and tents out in the harsh mountain country where drinking water and food supplies had to be carted in daily by the A.W.C. trucks. There was no electricity, refrigeration or amenities to make life easier. The bore water out there was unfit for drinking and was so hard that when Mum wanted to do the washing she'd put soap powder into the copper, boil it up, skim off the scummy froth and then boil up the clothes again so they'd be reasonably clean.

When they were permitted to return to Alice Springs towards the end of the War, their house had been damaged quite extensively inside, trees cut down and the garden flattened and turned over into tent sites. It took years for them to receive even partial compensation for those losses. And through all of this they remained very civic-minded and loyal and showed no public display of bitterness and anger at what had happened. This is only a small indication of the fine characters of my parents.

Chapter 26
Out at the Salt Lakes

Another episode of my working days included the birth of the salt industry in the Centre. Around 1937 I had decided to establish a salt depot in Alice Springs to supply butchers and station owners with salt for domestic use and also salt licks for the stock. The salt lake was about 180 miles (290 km) south-west of Alice Springs near Erldunda. I drove my Mulga Express Mark II out to the lakes and initially we collected the salt by using muscle-power and shovels!

I returned again around 1945 with the Mulga Express Mark III and had to make a track across and around the salty flats and red sandhills for about 10 miles (16 km) from the main road. I had hoped to bring back about ten tons in the first load with the help of four aboriginal workers. We used axes and shovels to build a causeway over the boggy outer edge of the lake.

We had only completed half of the causeway by Day Three when it started to rain and continued to rain all night. Next morning, Day Four, we loaded about three tons of salt using bags to cart it out to the truck. Everything was wet and boggy and it was still raining when we packed up our gear and headed for home in the morning of Day Five. We became stuck in our first bog after travelling only about one kilometre. When the rain eased we dug out of the bog went another half a kilometre and struck bog number two! It began raining again and we were all soaked while trying to dig out of there. Everything was soaked – even the matches. We made camp on a sandhill with the cover of our swags strung up over the bushes. Using some petrol out of the tank I soaked some rag and ignited it from the spark of the ignition wire. Luckily there was plenty of wood around so we had a big fire.

Next morning, Day Six, the rain had eased but there was water everywhere. We had only brought food and water for about a week, plus a .22 rifle and by this stage we only had a small amount of tea and flour left. We shot a couple of kangaroos on a nearby sandhill for food – one small and one medium sized which provided a good meat supply. Of course we had plenty of water! We kept the fire going and stayed there all day, cooking our meat, making damper and drying everything out.

Day Seven we headed off again, getting in and out of several bogs. One of the boggy holes was such heavy going that the tail shaft twisted off on 'Bitzer' Mark III, the six-wheeler. I didn't quite know what to do, but I rummaged around and found a piece of pipe on board that just fitted outside the broken tube of the tail shaft. But how to fix it? No welder, of course.

I searched my toolbox again and found a small chisel, about 3/8th inch, (10 mm) wide, so I sat down with a hammer and chiselled eight holes through the pipe and then eight holes through the tail shaft, which took about eight hours! I used some bolts out of the bodywork of the tray of the truck and bolted it all up. By the next day things had dried up a little more and we finally made it through to Alice Springs. The tail shaft vibrated a bit because it wasn't quite straight, but it was better than walking many miles for help!

When I started building the road trains I didn't have time to look after the salt supplying business so I gave the lease and the customers to Jack Swanson, an old pensioner, who wanted to earn a bit of extra money. He used his old truck and continued out there off and on for many years. About 1970 Jack took on a contract to supply 500 tons of salt to the O.D.E. oil drilling company, which was a job much too big for him to handle alone. He asked if I would do it, as the salt had to be delivered within six weeks. I constructed a salt harvester, finished building the causeway onto the lake and delivered the salt in 50-ton loads with my road trains within five weeks.

The salt lake industry has now been taken over by David ('Freddo') Frederiksen from the Solar Ponds project in Alice Springs which was using 900 tonnes of dry salt for laying a 1-metre deep base for the concentrated brine in the ponds. Freddo has also supplied Peko Mines at Tennant Creek with thousands of tons of salt and other Northern Territory salt requirements, including Alice Springs swimming pools and butchers.

The salt harvester I constructed in 1970.

PART IV
POST-WAR YEARS (1945–1951)

CHAPTER 27
BUILDING ROAD TRAINS

At the end of the war, back in Alice Springs, I started a garage and worked at carting, general engineering and motor repairs. I still had my six-wheel truck, 'Bitzer' Mulga Express Mark III there, built from some components of my original Willys-Studebaker-Buick-Dodge vehicle which I had used on the mail runs before the war.

Des Byrnes, who was courting my sister, Mona, joined forces with me in acquiring the General Motors agency which we ran as Northern Motors for about twelve months. Des was our bookkeeper and accountant; he and Mona were married later. When the Army disposal sales began we bought quite a few ex-Army trucks and cars and reconditioned them for sale. Since new vehicles were difficult to obtain our reconditioned vehicles were in big demand. I later sold out my interest in the agency.

I had already been involved in general carting since I commenced working in Alice Springs at the age of fourteen, but my work in cattle transport actually started after World War II when I began carting livestock and other general carrying, which led to building my road trains.

Stud Bulls to Elkedra

Bill Reilly from Elkedra Station had some expensive stud bulls coming up on the train and asked me if I could put a crate on my truck and cart the bulls up to Elkedra as it was 300 miles (480 km) to walk them there. He asked, "Now, are you sure you can do that?"

I answered, "Yes, I reckon they'll travel all right."

This was accomplished successfully and the cattle arrived at his station in really top nick. As a matter of fact, they were so healthy when they arrived at Elkedra Station in the evening, that after unloading them they went straight off the truck and onto the milking cows which were on their way into the yard! Old Bill commented, "Well, you certainly didn't do them any harm!"

About 10 miles before Elkedra, on the very rough track going into the station, a stub axle on the truck broke and gave me a lot of trouble replacing it, getting the old bits out. While working on that I gave my thumb a nasty bash. It already had an infected sore on it and that bash on my thumb-nail really made it flare up, so I stayed there a week with a poisoned hand. Molly Reilly, Bill's daughter, looked after it. She seemed to take great joy in putting red-hot poultices on it – they were so hot they even blistered part of it!

A Little Horse Tale

After successfully carting the stud bulls to Elkedra, word got around that I'd done a good job on that. Several people connected with racing came and asked how much I would charge to take some horses up to the Wauchope Race Meeting. I figured it out and said I could take nine horses for £90, or £10 each. They were very happy with that price, but there was some delay and argument about when I'd be leaving. Some wanted to leave three days ahead of the races and others wanted to leave a week ahead. There was also another argument with one owner refusing to have his horse carried in the same compartment as another person's horse; he made the accusation that the other horse was "a kicking bastard of a thing." The trip was cancelled until we came to an agreement at the last minute, a day before the races, with me carting six horses.

A police sergeant, known to be a bit nasty, was attending the meeting at Wauchope. Jim Maloney, the wooden-legged bookie from Tennant Creek was there too, running a book. Quite a few horses from the stations around had been entered for the races and there was a fairly big crowd.

The races were going along pretty well until another bookie and a couple of others tried to 'fix' a race. They'd paid £20 to one of the young jockeys who was riding the favourite horse (which I'd transported up), and instructed him to "hold the horse back." All the betting was in full swing just before the race started.

Jim Maloney knew this particular horse was the best horse, so he became suspicious when a lot of large bets were coming in for another horse with high odds. Jim smelled a rat and sent 'Brownie', his offsider, off to ask the young lad who was riding the favourite how much he was being paid to hold the horse. The young lad, inexperienced in wheeling and dealing, answered straight away, "£20." Brownie said, "Well, Maloney will pay you £40 to let him win."

Away the race went. Jim collected very well from that race, but those involved in the swindle, including the police sergeant, were absolutely ropable when they discovered they'd 'done their dough'. Their bribing of an innocent lad had come unstuck and they couldn't do a bloody thing about it!

Before the race a couple of chaps had come to me and advised, "Put your money on 'so-and-so' – she's all fixed!" Since I didn't know what had been going on and didn't intend to gamble anyway, I just said, "I'll keep my £60 I've received from the carting." And just as well I did!

Dick Turner, for whom I had carted some of the horses for the race meeting, had built and owned the pub up there at the time. When he heard of the scam he laughed his head off and said, "Serve the bastards right! I hope they lost plenty. That sort of thing gives the game a bad name!"

Army Disposal Sales

Ted Dixon, from Waite River Station had also seen the bulls being loaded for Elkedra Station. Word got around that this was the best way to shift the bulls. Ted asked me one day, "Why don't you try building a bloody big truck that would take a hundred head? I'd send all my cattle into market like that." I told him I already had plans made in my mind for building a road train with self-tracking trailers for negotiating bush tracks, but I didn't have the money to buy all the equipment which would probably cost about £10,000 just for starters.

When I saw him again later on he told me, "There's a lot of Army disposal stuff up for sale.

Maybe you can get enough junk out of that lot to build something. There's £2,000 here, any time you want it, free of interest, if that'll help you."

About three months later Army disposals had twenty-three Bren-gun carrier recovery trailers for sale. Those trailers were just a great big lump of a steel frame welded together for recovering broken-down Bren-gun carriers. They had two big wheels and stub-axles welded onto the frame and a spare tyre and wheel, all with big almost brand-new 1125 x 20 tyres on them. Since they were on offer for sale, I made an offer of £35 each. They 'ummed' and 'ahhed' for a while, because each tyre was worth about £75, but they eventually agreed.

I immediately chased around looking for Ted Dixon to take him up on his offer of the money, and found out he was attending the races at Barrow Creek. I hopped in my car and zoomed the 180 miles (290 km) up there, met up with him and told him what was happening. It was the second day of the races and Ted was fairly well-primed at the time. He called to his off-sider, "Eh, Matt! Go over there and bring my saddlebag."

The lad went over and got the saddlebag and came back to where they were all sitting around under the trees. Ted pulled out his cheque book, wrote me out a cheque for £2,000 and said, "There you are, young fella. Go to it! Pay me back when you can, and I want to have the first cattle carted!"

(In those days £2,000 was a large sum of money. The equivalent monetary value would be in the vicinity of $100,000 today. Without that loan it would have been almost impossible for me to even consider starting on my dream of building a road train with a capacity for carting over 100 head of cattle). I said, "We'll need to have a road out there to your place."

At the time he had only a rough track going through gullies and gutters, about 50 miles (80 km) further than it needed to. He said, "Well, you stir that bloody government mob up and see if they'll cut a track through from Connors Well to Woodgreen Station, and I'll put a yard up at my boundary." All we wanted was a track cleared through, but the government officials said they wouldn't do it because a road train of the calibre I was building would never work. This was the off-hand reception my ideas first received from bureaucracy. Several years later, when the success of my road trains was proven the government did build beef roads.

Building a Crane and the First Trailers

So that I could construct the trailers I used a concrete slab, which had previously sported an Army hut on it, as my workshop area. I then built up a welding plant out of bits and pieces. My biggest problem was getting the wire for winding the transformer, but I eventually accomplished it. After the arc welder was completed I obtained an old chassis and engine from a 1934 Leyland model tip truck. It was bare – just the four wheels and the engine. I constructed a mobile crane out of it using steel and other bits and pieces which I had bought. This crane enabled me to transport, lift and cart the equipment I had bought from Army Disposals. It also earned me some extra money for wages for Carl Koerner, who did the bulk of my welding work. In addition, I hired it out at £1 per hour to other buyers who came to Alice Springs as I was the only one in town then with a crane.

The crane I built up and used to construct my trailers for the road trains

After designing a rough sketch on paper, I started building the first trailers and these units formed my first road train. I removed the stub axles from the Bren-Gun trailers which had simply been welded to the frames of the trailers. I then made up straight-through axles for the trailers and welded the stubs to them. I made up bogies, turntables, tow-bars and inter-coupling chassis and crates. Then, of course, I needed a prime mover which was powerful enough to haul three loaded trailers – a road train 50 metres long.

I had ideas of building my own prime mover with a round nose similar to the modern big coaches, 36 feet long with self-tracking double bogies, with the back half as loading space and the other half of the unit as a power house and sleeping and driving cabin. I'd bought ten heavy-duty International rear axles and thought I'd make it 8-wheel drive (four axles) something like the old A.E.C. road train, but much longer and considerably more powerful. I planned to power it with twin-six G.M. diesels; in fact I tried to get a twin marine power pack but was unsuccessful in my efforts.

I sent away to U.S.A. for a quote on a G.M. 671 Twin diesel engine with a torque converter on it, which would have cost about £12,000. I organised with the Development Bank so that I could finance it, but when I provided details and specifications to General Motors about the unit they refused to supply a Twin. They insisted I'd have to buy a Quad 6, which was a 24-cylinder because they reckoned a Twin would be over-loaded. They, of course, were thinking of higher speeds. I had to abandon my idea because the Quad was around £25,000, which was a fortune in those days. So I had only my second trailer and was pretty much at my wits end wondering what to do for a prime mover, since I was out of money.

Money Matters

Here lies another story. I went to the manager of the Commonwealth Bank, Virgil King, and explained to him what I wanted to accomplish. He figured out that I needed about £2,000 to finish off the trailers. Since he didn't know much about their new finance branch he said he would write a letter to the Governor of the Commonwealth Bank in Sydney, which he did. He corresponded back and forth several times with them but they still hadn't made any decision about the loan to me and indicated that they wanted more information.

Virgil was getting a bit tired of it and said to me, "Here is a pad and pen. You write it out in your own words and you will probably receive an answer back straight away."

I said to him, "You know I couldn't write to the Governor of the Bank!"

He replied, "Of course you can. Your grammar is good and you understand your concept and can write it the way it should be written. The Governor of the Bank can't write a letter – he can't even read his own writing! He can't drive a car – he can't do anything but be a good banker! I'm the same," he continued as he laid his hand on half a ledger, "and that is all I can show for a year's work. I would give my right hand to be able to create something like you are doing. You would make us look a bigger fool if you asked us to do the sort of thing you are accomplishing."

I scribbled out a letter and he posted it off to Sydney and within a week I had my loan. He pointed out that the things you can't do yourself you pay someone else to do. Up until that time I had quite an inferiority complex because of my lack of formal education. That was a turning point in my life. From then on I was never embarrassed about my spelling or doing paper work.

Acquiring my Prime Movers

That money enabled me to finish my trailers. I had one completed, a second almost completed and a third trailer started. At this point I was becoming rather frustrated about General Motors refusing to supply me with a reasonable sized motor. Then I heard there was a big Disposals sale in Darwin and I was told there were some big diesel engines (like boat engines), big trucks and so forth, so I thought I'd go and have a look. All I had was my return ticket to Darwin on Guinea Airways and £80 in the bank. I needed to take a bit of a break anyway and wanted to have a look at post-war Darwin, so I travelled up and lo and behold! there was a beautiful Diamond-T tank transporter (they're like a tractor) with a 24-wheel low loader behind it.

The day of the sale many trucks were available, the Diamond-T being the last truck being offered. I was rapt with it because it had only done about 14,000 miles (25,000 km). I started bidding at £1,000 but it got up to about £1,500. I thought, "Well, blow it! I'm going to have a go and worry about raising the money afterwards , or get someone to buy it for me." My last bid was £2,250 and we were told, "Sorry, gentlemen, there's a reserve price on it for £2,500. We'll deal with the last bidder after the sale."

Then the battle to get enough money to pay for it started! I had four days to raise the money. I didn't quite know what to do. I tried a couple of people there who were well cashed up, but they didn't want to be in it. I then went to Neville Bell, a friend of mine, who was an agent for Custom

Credit. He said he'd need a third down for new vehicles and half down for second-hand vehicles. He could organise that part of it, but I'd have to dig up the balance.

Don Speed, who was a big buyer there at the sale, also had a finance company in Melbourne. He was also a real 'sharpy', but I didn't know that at first. He'd apparently been involved in some shady deals and had even 'done time'. He came to me and said, "I believe you bought that truck."

I said, "Well, I bid for it but I don't have enough money to pay for it yet." Of course, he knew very well what the score was, because he'd offered the auctioneers £1,500 for it if I couldn't raise the money. He offered, "Well, look, I can loan you the money." He immediately called over to his big, tall blond bookkeeper-cum-accountant and said, "Write out a cheque for Kurt for £2,000."

I said, "Hold on a minute! I want to know what conditions apply, what security you need, and what the interest is."

He replied, "Oh, don't worry about that. Without security, I'll just make it a straight-out loan at 33.3 per cent interest, and it'll all be fixed. You can take delivery tomorrow."

Doug Wheelhouse, another buyer whom I had helped out a few times, was within earshot and overheard what was going on. He called me over and warned, "Don't have anything to do with him! He's the greatest shark of all times!" I commented, "Well, it sounds a bit like it at 33.3 per cent interest." (That was in 1945 when top interest was around 10 per cent).

Building the first three self-tracking trailers

Doug said, "Look, I didn't buy as much gear as I thought I would this trip. Is £1,000 any good to you?"

I said, "Golly, yes! But what about interest and how long can I have it for?"

He answered, "Well, look – you've helped me out in the Alice with your crane and other things quite a few times before. You can have the £1,000 as long as I get it back by the next sale."

I asked him when the next sale was on and he said, "Your guess is as good as mine – it might be three months or it might be twelve months." Anyhow, I said, "Right. I'll take it on those terms."

The first Diamond-T prime mover as purchased and the completed first road train

So he dived into his big overalls pocket, pulled out a great roll of notes that a kangaroo couldn't jump over and peeled off £1,000 and simply handed it to me. I wrote him out a note on a piece of paper saying, "This is to certify that I owe Doug Wheelhouse £1,000."

I then went around to Neville Bell's place and said to him, "Well, I've got the deposit."

He said, "Right! We'll fix it up." So we signed the papers, obtained the details from the auctioneer of the engine numbers and so forth, and I set off with my Diamond-T Number One, Model 980.

I had managed to acquire quite a bit of back-loading and I'd also bought a 4-wheel, 10-ton trailer for £100. I loaded all Doug Wheelhouse's gear plus some for Eddie Hall and more for Alcoota Station, all pre-paid loading, which gave me enough money to pay for the trailer and a few extra pounds for fuel.

About eight weeks later, back in the Alice, I was busy pulling the truck to bits, lengthening it out and had almost finished it. Next thing another Disposals sale in Darwin was announced with another Diamond-T prime mover available, exactly the same model as the one I'd already purchased. I thought it would be really perfect if I could get that one too. I tried to book a ticket on a plane to fly up there straight away, but all the seats were booked out by southern buyers who were heading up there too, so I couldn't get there until the day after the sale.

By the time I arrived in Darwin I was disappointed to find that the other Diamond-T had already been sold. However, it turned out that Doug Wheelhouse had bought it for £3,500, which included a 24-wheel low-loader. Doug said, "I don't really want it, but I knew you wanted it so I bought it. But I'll keep it if you don't want it."

That meant I had to dig up £4,500 if I wanted the truck. Well, I didn't even have the money for my fare back home! I had bought a one-way ticket and I had about £20 in my pocket. I was virtually broke.

The sale was still in progress, but the trucks had been sold and they were starting to sell the other gear. Ly Underdown had bought a great heap of reinforcement steel which he wanted to use to build his new pub. There was supposed to be an estimated 170 tons of steel there. He approached me and asked if I had my truck up there. He told me he'd bought the 170 tons of steel and wanted me to cart it down for him.

I thought, "Hooray! (the normal price was £25 a ton for cartage, but for a big lot like that I thought £20 a ton would do, so I was counting my chickens already). I said to him, "Well, I'll cart it cheap for you at £20 a ton."

"No way!" he said. "I'll give you £10 a ton."

I answered, "Hell no! No way! That would barely pay for fuel and tyres."

He just said, "Well, the money is there if you want to cart it."

I checked around a bit to find more back-loading. Eddie Hall from Cummings was up there again and he needed about 50 tons of tractors and tractor parts carted. He agreed to pay me cash before delivery at £16 a ton, saying "That's the best price you'll get."

I said, "Right! I'll have that." That amounted to just over £800 including the tractors, which were a separate price. Then I got about another 50 or 60 tons for somebody else. It was getting close to what I needed and I worked out that if I got about £2,000 from Ly Underdown, plus a bit of extra loading that was kicking around, I could just about make the grade.

I went and saw Ly again and said, "Look, I'll cart if for £12 a ton if you can have the money here by Monday." By this time Doug was getting a bit 'toey' and wanted to finalise our business.

Ly said, "Right. I'll have it here by Monday – £12 a ton."

It turned out in the end it wasn't 170 tons – it was 210 tons! But I thought at the time that it sometimes pays to lose money if you get what you're looking for, because those two trucks were certainly bargains at that price. It gave me two identical models.

When I brought the couple of loads of steel down for Ly on this low-loader, old Mrs Underdown wanted to count it out bar by bar and measure the different lengths of bars, to make sure she got her right tonnage. Well, that was almost impossible, but I said, "I'll pull the truck up out there."

She told her son, Johnny, "You go and see that it is checked out properly."

I agreed, tongue-in-cheek, "Yes, Johnny can count it!" So I pulled the low-loader around into their back yard alongside the pub, unhooked it, put the winch off the truck onto it and just tipped the load of steel onto the ground, saying, "Right! You can measure all the thousands of bars and different lengths and count them!" (Of course, this was never done). I'd already put it over the weighbridge, but she wouldn't accept those weights. I got diddled on the weights, but at least I was able to get started on the road trains. I now had my two Diamond T's. Critics have said that I got started with my first road trains 'on the cheap', but in actual fact, in today's monetary values my first road train cost me the equivalent of $350,000 to $400,000.

The low-loader, which was a 24-wheel army tank transporter about 24 feet long, got cut and shut. I cut it through the centre and extended it with four 30-foot steel girders into its belly and extended it both ends, making it a 75-footer so that I could move some big buildings around south of Darwin and later used it for other contracts.

Transporting a house from an Air Force base to Katherine, N.T.

Chapter 28
Early Cattle Transport

Beginnings 1946 to 1950
The first consignments of cattle using my trailers were from Murray Downs to Alice Springs, a distance of about 230 miles (370 km). My cattle trailers were also able to be converted to flat tops for carrying general freight. These self-tracking trailers, which I made up in 1947, were 43 feet (13 metres) long and weighed 8 tons. They worked perfectly up to a speed of about 25 miles per hour, but weren't practical for higher speeds. With a prime mover and a string of three trailers 54 metres long it was possible to drive off the street into a standard-width gate without any worries at all. The trailers which were specifically designed to follow the prime mover tracks perfectly, were ideal for transporting cattle on the narrow, unmade, winding, sandy bush tracks.

I had extended each Diamond-T prime mover from 9-foot (2.7 m) to 24-foot (7.3 m) body trucks. My trailer fleet gradually increased to eight units, most of which had cattle crates and were four-axle, using big single tyres. I ran the wheel bearings in oil instead of hub grease which proved to be very successful. When my trailers reached the end of commercial life, having travelled almost 2,000,000 miles, the wheel bearings were as good as the day they were first fitted, in spite of having done most of their work on dusty corrugated tracks.

The second consignment of cattle using my road trains was for my friend Ted Dixon who had first encouraged me to get started in the cattle transport. This lot were transported from the Waite River Station boundary to the railhead at Alice Springs. There was also an early test run from Anthony Lagoon to Alice Springs (610 miles or 980 km) done for the benefit of the Departments of Transport and Veterinary Affairs with their livestock experts from Canberra and some of the other States, checking to see how viable this form of transport was and what possible

My first load of cattle from Murray Downs to Alice Springs, 1946

effects road train operations might have on road surfaces in the Territory. Ted Pettit, my first driver, had also helped me in construction of the road trains. In the early days, 1947-1948, the station owners had to be persuaded and convinced that transporting cattle by road, being faster, was better for their livestock than droving, even though it was more costly. If the cattle arrived in the saleyards in better condition they would, of course, fetch higher prices which would help pay for the road transport costs. In addition they could sell off much younger cattle (around two year olds) which weren't able to walk long distances without losing a lot of condition. Young cattle are what the market wants.

Unloading the first load of cattle using my mobile loading/unloading ramp
Alice Springs, 1946

The beef roads scheme hadn't even been thought of at that time. Most of my vehicles had to use rough bush tracks, other than the Stuart and Barkly Highways. After a few trips over them by the heavy multi-wheeled road trains, the roads compacted fairly quickly after light rain but heavy rains caused a lot of problems with trailers having to be pulled out one by one.

Around 1950 the Railways jacked up their price for cattle transport to Adelaide by about fifty per cent because of higher costs plus having previously been highly subsidized. Also the Northern Territory cattlemen at that time weren't paying income tax and the Government had decided they could well afford to pay a bit more for shipping their cattle.

I was having a similar problem to the Railways myself, with rising costs for fuel, tyres and wages! I increased my price by ten per cent, which created a storm of protest among the cattlemen – they virtually 'blackballed' me. A meeting of the Stock Owners' Association was held and they voted to keep using me if I didn't raise my price. I just said to them, "Well, sorry, I'll have to cancel the bookings." I had about fourteen train loads booked at the time, but I insisted, "If I can't get the extra ten per cent I can't carry on."

They refused to pay the extra so I stopped using road trains for cattle transport. I simply said to them, "I'll put my trucks in the yard. They don't cost anything standing in the yard, but they would cost a lot running them at a loss over the rough tracks."

In the next decade the Government actually graded a lot more of the tracks, but in the meantime I turned to mining and carting ore over to Mount Isa. Later, when the drought was very severe, several truckies with small semi-trailers started to carry the starving stock into the Alice and sending them down south where there was a good market. The cattle owners were very lucky at the time because in the south-east of South Australia and also in Victoria there had been very good seasons and they were buying breeding stock at very good prices. This also meant that they were willing to pay a reasonable amount to have their stock carted. I returned to cattle transport in the late 1950's which is described in a later chapter.

'Bertha', one of my first road trains being inspected by Aub Melrose (right) in 1965, with a view to unifying road transport standards, speeds, etc. for the Commonwealth

Chapter 29

The Famous 'Drum Case'

Around 1948 Commonwealth Disposals called tenders for all the surplus drums which had been used up in the Territory during the War. Jim Bowes, Chief of Disposals for Australia at the time, had mentioned the drums to me two or three months earlier. By a bit of good luck I bumped into him again in Melbourne just as I was heading off to the airport to return to Adelaide on a late afternoon flight. I'd been down there to organize the construction of two more trailers (built to my specifications by McGrath's) for my road trains. Bowes said, "I thought you were going to tender for those drums I told you about."

I said, "Oh, I was, but you can't be in everything. I've been really busy and I'd forgotten all about them."

"Well," he said, "They're going for nothing – going for absolutely bloody nothing! The tenders close at ten o'clock tomorrow morning, so you better get smart if you want to get into it."

I walked on up to the T.A.A. terminal and sat there waiting since I had about a half hour wait for the bus. Ideas starting rumbling around inside my head and I thought, "Well, I might as well have a go. And I wonder what he means by 'for nothing'? That could mean ten shillings each or anything!" I was up to my ears in debt already and I couldn't lose anything by making a bid.

I decided I'd investigate and worked out how much money I had left. I figured that I could afford to tender at four pence halfpenny each so I sat down and wrote him a telegram which said, "I hereby tender four pence halfpenny each for all the drums in Darwin" and sent it off at the airport office before boarding my flight to Adelaide.

Next morning at about quarter past eight the phone rang with a call from the Disposals Commission Tender Board in Adelaide. I was staying at the Imperial Hotel which was only about three or four doors away from the Disposals Tender Board. They said, "We have this telegram here stating that you have tendered for the drums in the Northern Territory. Will you come over and fill out an official form? You must be here before ten o'clock."

I said, "Okay!" and immediately did that.

Then at about nine o'clock I received another phone call from Melbourne from the Shell Company saying, "I believe you are the successful tenderer for the 40-gallon drums up in the Territory. If you haven't made any prior arrangements, we would like you to give us first offer on all the good drums."

I found out later that Shell Company had a 'stooge' in the Disposals Commission who had let them know what the highest tenders were. It turned out that Caltex had tendered at threepence a drum, Mobil or Vacuum Oil Company had tendered at threepence halfpenny a drum and Shell Company had tendered at fourpence a drum. My tender, of course, was fourpence halfpenny. Mr Angel, who was the Chief of the Shell Company for Australia at the time, told the bloke in the Adelaide Shell Company Office to ask me to come and see him in Melbourne. He commented,

"Well, you wouldn't bloody read about it, would you? A bloody mug from the Bush knocks us out! We thought we had it all tied up nicely."

When I went into the Disposals they asked, "Are you quite sure you are satisfied with the tender?"

I replied, "Well, how many drums are in it?" They told me there were 63,000 altogether and there were two and a half tons of bungs at the Shell Company Depot in Darwin.

He asked, "Now, you are quite sure you don't want to cut it down?" The cost was about £1,180 with the bungs. I wasn't quite sure then and showed some hesitation. I didn't realize it then, but he was trying to get me to push my price down. If I'd dropped it down below Shell's tendered price then they would have got the drums. I thought about it for a moment and answered, "No, I'm quite happy about it." So I got the drums. If I'd had £3,000 to spare I could have tendered one shilling per drum and still been happy with the deal!

Some of the drums had been used to carry fuel, others bitumen cutbacks, but they all needed to be sorted, cleaned and picked up from the numerous old Army and Air Force dumps. We picked through and sorted out the A grade drums first. Shell Company paid me £1 each for A grade and fifteen shillings each for B grade, delivered to Alice Springs. C grade drums were those which were slightly marked and needed cleaning. All I was interested in at first was to get my money back as quickly as possible. I began carting them 1,150 at a time using three trailers and a body truck. At first I sold Shell about 3,000 drums straight off, which paid for the drums and most of the cartage. Eventually, after most of them had been sorted through and sold to the different companies, I had about 10,000 of the drums which had been used for carting bitumen cutbacks left, which had to be cleaned. I set up cleaning depots both in Alice Springs and in Darwin and built a couple of tumbling machines for cleaning them, using chains and kerosene. I employed Jimmy Hodgman as my foreman and chief worker. He was the best worker I ever had. Anyhow, he was in charge of the Darwin end of things – sorting, cleaning, stacking and marking the

Some of the 63,000 drums I transported from the Northern Territory

drums. I had two road trains operating then, but mostly used only one carting drums. When a road train arrived up there Jimmy organised the crew of three drivers and his off-sider to load it up and sent the drums off again.

We had just about reached the bottom of the barrel, but were still sorting down at Brocks Creek, Pell and a few other air strips. I said to Jim, "You'd better come down for Christmas, with the next load." Jimmy had already marked and stacked up around 8,000 drums ready for loading.

The day after Christmas I received a phone call from Clive Graham, the Superintendent of Police up in Darwin. He said, "You'd better get back up here as fast as you can if you want any of your drums. Somebody is carting them out and stacking them."

I immediately up and off back to Darwin. It appeared that quite a few of the pilfered drums were being sold to Caltex by 'Bogger' Young who was paying his truck drivers ten shillings a drum for any good drums they brought in for him on their return trips to Darwin with empty trucks. He had quite a few tip trucks and a couple of semi-trailers and Bogger's drivers would throw on 100 here or 200 there or 50 on the smaller trucks. They were stacking the drums on a vacant block in the Darwin area and selling them to Caltex. Of course, the drums had all been nicely marked 'A', 'B' and 'C' by us.

Clive Graham had seen them coming in and realized where the drums must be coming from, so he kept watch to see what was happening to them. When I arrived in Darwin and went to discuss it, he said to me, "Well, look – I'll make it simple. We'll charge him for false possession."

The police then confiscated the 8,000 drums Young had stacked and had them carted into a vacant block alongside the Police Station. This led to the famous 'Drum Case' which lasted for four months in the Lower Court. I was a Crown witness and Young was charged with 'false possession.' The case went on and on and all sorts of 'red herrings' were dragged across the track. The lawyers really had a field day. Actually, I was lucky I didn't have a lawyer – I had the Crown fighting the case for me.

Young alleged that they were his drums and hadn't come from my areas. We had to prove that Commonwealth Disposals hadn't sold Young or any other person any drums – they had all been sold to me. Young made sure that we didn't subpoena any of his drivers as witnesses to say from where they had obtained and loaded the drums. He did this by paying off all the drivers involved and he told them to "get lost" interstate without leaving any forwarding addresses "or they would be in big trouble."

Young's lawyer was Leo Lyons, a very bullish type of man, who ranted and raved in the Court. He obtained Jerry Fanning to help substantiate their case. Fanning was also involved in the background (we later found out). Many years earlier before coming to Darwin, Fanning had lived in Western Australia where he had trained as an industrial chemist. They got him to swear that the grease crayons we had used to mark the drums could not last in the hot conditions and climate up there. The Defence was attempting to prove that the drums which had been marked three months previously by Jimmy Hodgman would have had their marks completely obliterated by then.

Fanning's evidence was given in the first week of the hearing. The Defence claimed that the grease crayon marks on the drums had been put on by their men. Of course, Fanning didn't know that the case would drag on for six months, so the crayon marks they claimed that they had put on the drums were still bright and clean after six months! Of course, by that time the Court had also proven that the Commonwealth Disposals had sold all the drums from all the abandoned Army and Air Force bases to me only, and his claim that after three to four months nobody would be able to read the crayon marks made on the drums was obviously false.

The Lower Court was used for a lot of general run-of-the-mill cases and only sat on the 'Drum Case' about two or three days a week, which dragged it out considerably. Eventually, after six months and a lot of adjournments, the Magistrate summed up the case and fined Bogger Young all the Court costs against him, plus a sentence of six months in jail.

Young appealed to the Supreme Court. Six months later the Supreme Court held a hearing which lasted for about one month. Just before the Judge was ready to hand down his decision, he took ill and was flown down to Sydney where he died later. That left the case still swinging in the breeze. Those drums were starting to become an embarrassment to the Crown and they didn't quite know what to do about it. It took about another six months before another judge was imported. In the end the whole problem was solved when everybody, especially me, became thoroughly fed up. It had cost me a fair bit in lost time and business, although I was still operating down in the Alice with some of my crew, but my business was going downhill.

Around that time the Darwin City Council had decided to put on the first-ever Darwin Show. During the first day of the Show nearly the whole population of the town was at the Show, including the Police. One policeman was left in charge at the Police Station and he didn't hear anything that was going on next-door in the vacant lot.

Young organized all his trucks and drivers, and some labourers to shift the drums, and within about four hours they removed all the drums from the Police Station yard. It was quite a good effort as far as speed was concerned! He again claimed that those drums belonged to him – he was calling our bluff to start all over again.

It was up to the Police to do something so they phoned me and asked, "What do you want to do about it? Do you want to take the case on?" By this time I was thoroughly fed up with it and back down in the Alice working and keeping very busy, so I said, "No, I've had enough." So Young finished up with some of my profits.

By that time the value of the drums had risen. Since the end of the case I'd gone through the reject stacks, re-sorted and cleaned up something like 6,000 more drums which the petrol companies bought for 35 shillings a drum. They all had to be repaired, but the companies were so short of drums and the cost of new drums had risen so dramatically that they were quite happy to pay a good price for third-grade drums.

I encountered Young a couple of years later. He was a hale and hearty bloke, bold as brass, and even in the middle of the Court case he'd come over when I was in the Club to have a chat or sling some jokes from across the other side of the bar. When we met again he said, "Well, just as well the prices went up and I'm glad you didn't have a go at me for the rest of them. I couldn't have afforded it – I was broke!" He told me straight out that he'd been bluffing and it had worked. By the time he sold all the drums he came out just about square after paying his legal costs. So that ended the famous 'Drum Case' which was one of the longest-running cases in the history of the Territory up to that time.

I had another stack of the last 4,000 C grade drums in my cleaning yard and all needed repair or were a bit rusty and leaky. A chap named Favaro came to me and asked, "How much do you want for those drums?" I said, "Oh, you can have them all for £1 each."

He told me his trucks were returning empty to a customer at the meatworks over in Queensland. Anyway, he offered me fifteen shillings each for them which I agreed to accept but he hadn't handed over any money yet. I gave him the key to the yard where I was allowing a chap to live, acting as a sort of caretaker. Next thing I discovered the 4,000 drums had disappeared. Favaro claimed that he never took any of them and that the other bloke had left the gate open. However, Favaro also disappeared soon after and I found out later that Favaro had left the Territory and carted them all the way over to Queensland, so that was the end of another batch of ill-gotten drums where another crook cheated me out of some of my profits! It wasn't worth the time and money to chase after Favaro. Actually, I had done very well out of them and with the money from them I managed to finance the building of my second road train and later took on the Railway contract.

During that period I had been building the second road train and other trailers, constructing my workshops and doing some cattle carting. Besides building up my workshops I built a house in Undoolya Road, Alice Springs. At that time there were only about half a dozen dwellings over on the east side of Alice Springs, and most of them were left-over Army sheds. When we returned from Brisbane, Kath and I had at first lived in an ex-Army storage shed and gradually built up from there.

My Driving Crew

I had quite a good crew of men working for me during those years. Carl Koerner and Jimmy Hodgman were two of the foremen I employed and others of my crew included Tommy Denny and his brother, Bert Noske who married my youngest sister, Myrtle, Ted Pettit, Tom Ryan, Henry and Ted Kunoth. Other early drivers included Roger Wait, Peter Ritchie and Ivan Wiese. One snag was Bill Llewellyn. I caught him out 'flogging' some of my fuel up the road. Usually we'd carry enough fuel to see us through up near Darwin and back. When carting the drums, if there was a head wind on the way we needed a couple of extra drums of fuel.

One weekend I had already fuelled Llewellyn up for a trip and put the extra couple of drums on. Just before he left I caught him out that Sunday, when he thought I was down at the Club. He'd put on an extra four drums of fuel which I discovered he was flogging to his good old mate, Noel Healy, at Dunmarra.

He was also involved in the disappearance of some of my tyres. Since we always needed extra tyres on board, I'd sent up ten extra tyres, worth £120 each (or $5,000 worth) to Dunmarra, which was about the half-way point of our run, so they were available for my drivers to draw upon in case anyone had a bad run. Llewellyn came through in the early hours of the morning and dropped the ten tyres off at the side of Healy's pub. Healy knew they were coming since I had phoned him and asked him to store them for me. When the tyres arrived 'someone' carted them away and stashed them in a pumping shed down by one of Healy's bores. When I asked about them, Healy claimed that he'd never seen them, saying, "Some rotten bastard must have knocked them off!"

I took a trip up there in the ute and made a few inquiries but nobody seemed to know anything, so I decided to drive around and check things out for myself, which is how I discovered some fresh tracks leading to where the tyres were stashed. I recovered them, loaded them back into the ute and took them home with me. Of course, nobody at Dunmarra knew how they'd got into the shed!

Three of the fuel tank cars I carted for the Commonwealth Railways

Carting Railway Engines and Rolling Stock

My road trains were also used for carting railway engines and rolling stock on a contract I obtained from the Commonwealth Railways to move all the surplus rolling stock from Larrimah to Alice Springs, 625 miles. This rolling stock, weighing 2,000 tons, comprised 3 tankers, 150 goods vans, 3 engines and their tenders. That contract, which helped really put me on my feet in transporting, was completed on schedule in 156 days. There was only one near disaster when the road gave way on one side after heavy rains and one side bogged right down leaving 3 trailers at a 45 degree angle with the goods vans ready to topple over. As it happened, 3 other transports were following and came to the rescue and managed to pull the trailers back on the road by placing a cable to the top of each goods van and simultaneously pulling together sideways and forward.

That infamous 'Drum case' had perhaps cost more than a loss of money. I was away for long periods up in Darwin on the Court case and also working extremely hard in building up my road trains; in the meantime my marriage with Kath, which was already shaky, deteriorated further.

The Famous 'Drum Case'

One of the three locomotives and tender I transported from Larrimah to Alice Springs around 1949-1950 Each locomotive weighed 26 tons.

More rolling stock which I transported for the Commonwealth Railways
Top—one of the tenders, weighing 12 tons
Bottom—2 goods vans, 1 freezer van and 1 flat-top

CHAPTER 30
ACQUIRING WINGS AND CRASHING THINGS!

In 1948 my two road trains were busy carting cattle and Army Disposals gear from Darwin. Leaving Carl Koerner to look after operations, I decided to go south for a break, joining Kath who was already in Adelaide, which was shortly before we parted. The three boys – Lindsay, David and Peter – stayed with a Mrs Steer, whose son worked in the Railways, in Alice Springs.

While I was down south I decided to learn to fly an aeroplane so I took lessons for about three weeks at Parafield airport, fulfilling the yearning I'd had from childhood days. When I had acquired my licence, Pat Davis, a friend of mine, suggested I buy his Miles Hawk plane for £800, as he was buying a Percival Gull. I had about 50 hours of flying experience by the time I was ready to fly home with Kath.

Kath and I with my Miles Hawk aeroplane, before the crash, 1948

I took out insurance cover for the plane on the day we left. We'd been flying against a strong head wind and I was running about one hour late. Around dusk, as we approached Mount Eba for landing, I saw the airstrip beacon from about 20 miles away, so I didn't have any trouble finding the place. The trouble was I had no landing lights and there were no lights on the runway. Because of my inexperience I made a mess of the landing and crashed. The plane was about 6 feet off the ground and travelling at 70 miles an hour when the nose dipped, possibly due to a wind shear. The plane just seemed to drop and hit so hard that it shot up again and then took an almost vertical dive into the ground. The tail broke off, the motor was flung several feet away and the fuselage crumpled all around us.

Kath was unconscious and had sustained a broken leg and a bad cut across her jaw, while I finished up alongside the motor on the ground still strapped to the seat. The motor had broken away from the fuselage, while the instrument panel and the windscreen were draped over my head. I received only one little scratch on my finger! The dust hadn't even cleared when I freed myself from the seat straps and dragged Kath clear in case the plane caught fire.

A mangled mess at Mount Eba after crashing

An R.A.A.F. Dakota was sent up from Mallala to Mount Eba and arrived around 2 a.m. that night and flew us back to Adelaide. Wing Commander S. Dunstone, a senior R.A.A.F. medical officer in South Australia, had attended to us on Mount Eba field and ordered that we be returned to Adelaide. Kath was hospitalised for about a week and then continued as an out-patient for a while. I was examined but they couldn't even find a bruise on me. Flight-Lieutenant J. Archer of the Dakota crew, who had been in a few crashes himself, was absolutely amazed that I came out of the wreck without a scratch. He said he had never seen anyone walk away from a wreck that looked as bad as this one. Another R.A.A.F. crew member who saw the damaged plane commented, "He is the luckiest man I have seen for a long time."

A few months after returning to the Alice I purchased a Tiger Moth which had just been overhauled and test flown. I took it for a flight and everything seemed to be operating perfectly so I came in to land. As soon as I touched the ground the left wing dropped, so I went around again and the same thing happened. I took off again, by which time I realized there was something wrong with the under-carriage. I approached the landing strip again, this time as far over on the right side of the strip as I could. When I touched down I held the left wing up as long as possible. When it dropped onto the dirt strip it dragged along until it hit the windrow on the left side of the strip, dug in and flipped the plane upside down. In my panicky efforts to get out, while hanging upside down, I pulled the seat harness pin and fell out like a bag of spuds onto my head, nearly breaking my neck. When Mr Emerson from D.C.A. checked the plane he found a pin holding the left wheel of the under-carriage was missing.

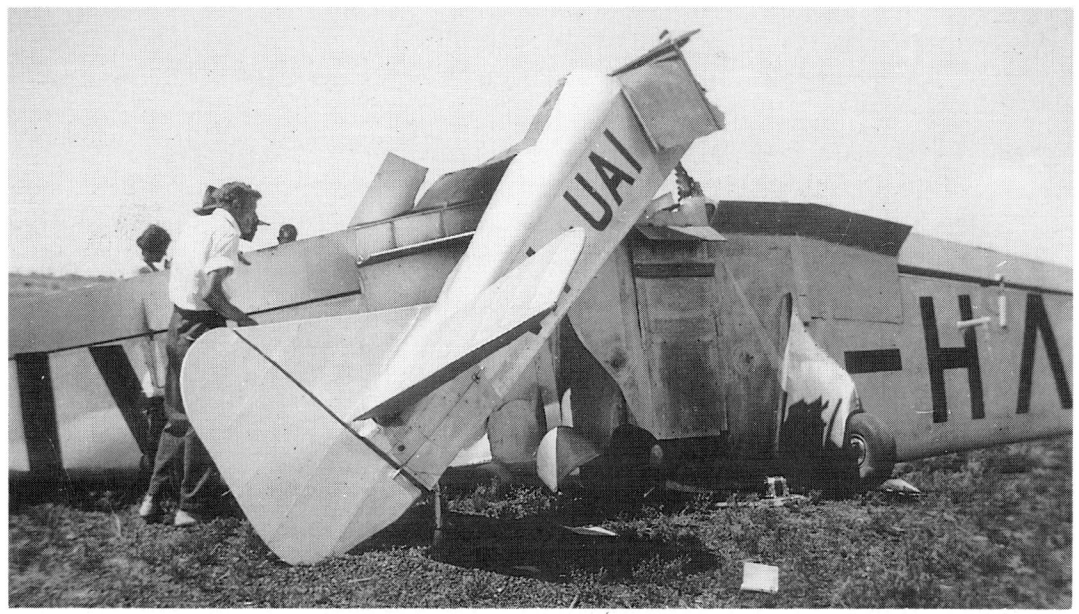

The wreck of my Miles Hawk aeroplane at Mount Eba

My next aeroplane, a Tiger Moth, 1948

I then bought another Tiger Moth and flew it for about 1,000 hours including my trip out to Lasseter's country, recounted in Chapter 35. I later sold it when I had no more use for a plane and the novelty had worn off. A few years later I tried some gliding but I gave it up as I needed to spend most of my time working at Jervois Mine and didn't really have time for such pleasures.

Kath and I divorced late in 1948 and no, I didn't deliberately crash the plane earlier that year! When she left Daphne Hillam had been installed by Kath to look after the three boys and also as housekeeper. At the time, Daphne was working at the Alice Springs Hospital as a dental nurse. When I phoned the hospital after arriving home, having not met her at that point, but knowing she was a dental nurse, I said in a muffled voice, "I'm from Borroloola and I've got a terrible toothache." She said she'd arrange an appointment for me right away. When I arrived at the hospital I announced, "I'm the man from Borroloola!" She, of course, immediately recognized who I really was from photos she'd seen at home and thought it a great joke. Kath returned after a few months and took David and Peter to live with her for several years in Darwin and Western Australia, while Lindsay remained with Daphne and me. We fell in love and married in 1950. David and Peter came back to live with us when Kath remarried Mick Greatorex.

Peter and David Johannsen and brother Lindsay extracting a witchetty grub from a tree.

Chapter 31
Prospecting at Bonya Hills

I first became involved in the Bonya Hills, just west of the Jervois Range, around 1949. Two Americans, Sharpe and Wright and their wives, from Queensland, had taken up a copper prospect there. They were hoping to achieve great things with it, but found it was beyond them as they didn't have enough mechanical knowledge. They talked me into buying their equipment which consisted of a diesel truck (a semi-trailer), some camping gear and also a small Ford Blitz truck.

At the time Jimmy Hodgman and Tom Ryan were working for me. I talked it over with Jimmy and he was willing to have a go at working in partnership with me. I was to supply the money, since I was still very busy with the Army Disposals work and cattle carting and Jimmy agreed to work it employing a couple more chaps, and we'd split the profits. It looked as if the ore could be concentrated and I had ideas of leaching the tailings and making copper sulphate. I didn't know very much about making copper sulphate at the time but I was going to have a try at it.

We had a small herd of goats out there. I flew my Tiger Moth out to Mount Swan to Bill and Elsa's place, and acquired a healthy young Saanen billy-goat to improve the quality of the herd for milk production. Young Martyn Petrick came back to Bonya with me. He sat in the front with his box camera at the ready for taking photographs and the billy-goat was hog-tied in the luggage compartment.

As I circled Bonya preparing to land I started to pull on the joy stick to slow the plane down. To my consternation it wouldn't go any further, so I had to land too fast – at about 80 miles per hour instead of the usual landing speed of about 60 miles per hour! It turned out that Martyn had put his camera into the joy stick cavity on the plane. However, we landed safely and the billy-goat went on to propagate the herd when he matured.

In hindsight, the whole Bonya operation was a disaster from the very beginning – it simply couldn't have ever worked. The ore body was too small, the grade was too low and only a small proportion of the ore was suitable for leaching. However, that was my introduction to the Bonya Hills area.

While we were there I did a bit of prospecting and came across a few small prospects of scheelite. On making inquiries as to the market, I found that scheelite was not in demand. Wolfram (a tungsten-iron-manganese ore) was needed, whereas scheelite is tungsten-calcium, which needs an electric furnace to smelt the tungsten out of the scheelite. Also, just after the war, neither wolfram nor scheelite was in much demand.

At any rate we closed the place down and I put it into the back of my mind and forgot about any ideas of developing it at that time. I left quite a bit of my machinery there and eventually around 1950, when copper prices were high, my brother-in-law Bert Noske had a go at it. He had been working out at Hatches Creek on wolfram when prices were high. Unfortunately he got "dusted" and contracted silicosis out at Hatches but it didn't show up as a health problem until he went out to Bonya. This ended a second effort at getting it going.

Chapter 32
The 'Case' of the Missing Gold

A few years after the War, Nobles Nob Gold Mine at Tennant Creek was losing a fair amount of gold, but the owners could never prove where it went or how it was stolen. All they knew was that every now and then some gold was swiped from the amalgam tables and instead of obtaining an average crushing of about two ounces to the ton, they'd run along for about eight hours and hardly get any gold off, or they'd get the gold off but it would suddenly disappear.

The culprits were never caught but it was 'known' that some miners had worked out some pretty shrewd moves to 'lift' it. After the gold was stolen it was put in with one of their mates' crushings in another mine, which would then show a very rich crushing, although the ore which went into the Government battery didn't have a very high gold content.

Another time Sandy McNab, a detective from Darwin, and a detective from Adelaide were running around up there trying to find a large consignment of gold which had gone missing, but they had no luck in tracing it. Sandy McNab and his mate had given up trying to track down the missing gold and were heading back to Adelaide when they called in at the Home of Bullion Mine to have a look around.

Toby Becker, who had been up in the Territory for many years connected with mining, had a small prospect of a mine near Tennant Creek. Besides that he was also working the Home of Bullion Copper Mine. The two detectives suspected that Toby Becker might have had something to do with some of the missing gold or at least know something.

Toby was on fairly friendly terms with McNab, so when they called in he invited the two detectives to stay the night since it was fairly late in the evening when they arrived. When they mentioned they were heading back to Adelaide Toby asked, "Oh, could you do us a favour? I've got a little battery charger engine that's packed up and I want to send it down to Adelaide to get it fixed ." Sandy obligingly said, "Oh yes, no worries."

Next morning Sandy put the little battery charger into the boot of the police car. It had the delivery address on it and Sandy promised to take care of it for Toby. Actually, Sandy himself was only going as far as Alice Springs, but the other detective was returning to Adelaide. Little did they know that there was about 60 or 70 ounces of gold amalgam stuffed into that little engine! Toby had taken the cylinder head off, put the amalgam inside the motor and closed it up again, so the two detectives were an unsuspecting party to the gold thieving! It was delivered safely to the 'fence' in Adelaide, which was quite a joke for Toby when he mentioned it to McNab years later.

About ten years later when Sandy McNab was stationed in Darwin, Toby and I met him in the pub. Toby was connected with me in opening up the Rum Jungle Uranium Mine. We were having a few beers and Toby conversationally asked McNab, "You know that gold that was going off in Tennant Creek – did you ever catch the culprits?"

McNab answered, "No, the bastards were too smart for us. We had a few suspects in mind but they only led us up a blind alley."

Toby continued, "Well, you know that little engine you took down for me.....?"

"Oh yes," McNab replied, "That's right – when I called in at the Bullion."

Toby continued, "Yeah.....Well, that was filled up with gold, and you bloody well delivered it for me! A poor bloody detective you turned out to be!"

Well, McNab's face was a picture! Of course he couldn't do anything about it by then. It was one of many incidents which occurred in Tennant Creek.

CHAPTER 33
LUCKY ESCAPES AT RUM JUNGLE

In 1950, shortly after I stopped cattle carting the first time, I intended to start drilling at Jervois Copper Mine which hadn't previously been drilled. I organized a syndicate, Northern Drillers and Company, with Toby Becker as partner, to obtain a Diamond drill and finance the drilling programme. Since I had mining leases at Jervois, I agreed to peg a lease known as 'Hanlon's Extended' in the name of Northern Drillers and Company and transfer some of my leases to the company as they were drilled. We decided Toby should go to Sydney to obtain the drill and arrange finance.

On his way back, Toby stopped over in Melbourne and spoke to Sir Maurice Mawby who was, at that time, the chief of the Bureau of Mineral Resources. After some discussions, instead of the drill going to Jervois, they offered our company a contract to open up Rum Jungle Uranium Mine, which was the first large uranium mine opened in Australia.

We accepted the contract and diverted all the gear which we'd intended to send out to Jervois, added all the mining gear from Jervois and some more from the Home of Bullion Copper Mine in which we also had an interest. Straight after the Wet we took a road train load of the gear up to Rum Jungle, about 90 miles (145 km) south of Darwin.

The Legend of how 'Rum Jungle' got its name
The story has it that, back around the turn of the century, the name for 'Rum Jungle' railway siding originated during the original construction of the railway from Darwin to Pine Creek gold fields. A goods van containing stores and some barrels of rum was left at the siding. Some of the local prospectors and miners camping nearby were having a booze up and ran out of grog. Next morning a chap went down to the gutter which ran by his campsite to get a pannikin of water to quench his thirst. Much to his surprise he found it tasted very strongly of rum. At first this bloke thought he was suffering from the 'DT's' and called out to his mate to come and have a taste. His mate also agreed it really did taste like rum.

The story goes that the gang working on the line had used an auger bit and drilled several holes right through the bottom of the railway van until they 'struck oil' (or rum). When the rum started to flow they filled their assorted containers and went off, letting the rum continue to trickle out steadily and flow down into the gutter!

Our Campsite
Our site was about 8 miles (13 km) north-east of Rum Jungle siding. Toby Becker, Rolly Ryan (a mining engineer) and I went out to cut a track in there, following a bit of a jeep track. We ran into some extremely swampy areas since it was just after the Wet season. Everything was wet, soggy and steamy, with mosquitoes which just about carried us away! The speargrass was 9 feet (3 metres) high, hanging everywhere and full of pollen which made us sneeze our heads off. To put it mildly, it was very, very trying, sticky and hot.

After nearly a week of continually getting bogged and making countless crossings over creeks, we managed to get all the gear in and set up camp at Coomalie Creek. On our second trip, when we returned with more camping gear, my brother, Randle and my wife, Daphne, also came up. This time we brought up a converted coach and used it as a caravan at our Coomalie Creek site.

Things progressed reasonably well for the first two months and we drilled in quite a few areas. The Government wanted a new shaft sunk and another old copper mine shaft, which was full of water, cleaned out.

Toby and the Scorpion

After we'd set up camp everybody was hot and sweaty and badly in need of a cleanup. We walked down to the creek through the high speargrass and stripped off, dropping our clothes in the grass on the bank. We dived in and had a lovely hour splashing around and cleaning up. By this time the sun had gone down and it was starting to cool off. We hopped out of the water and pulled on our shorts. Toby let out a helluva scream and ripped his shorts off in two pieces and found a scorpion firmly attached to his vital organ.

Next morning we asked Toby, "Er, how……..is it going?"

Toby replied, "If it wasn't so bloody painful I wouldn't have minded as it's the best erection I've ever had!"

Mozzies and the 'Citronella'

On our way up north to Top Springs on the Murran-ji track to fix one of the trucks, Daphne, Lindsay and I stopped at Elliott for the night. Elliott was notorious for its plentiful mosquito population. In those days the only available accommodation was a shop-cum-boarding house which had a spare room and shower on the verandah – no screening or protection from insects.

Once the lights were out the mozzies zoomed in for a feast, so in desperation Daphne finally got up and grabbed what she thought was the citronella to keep them away. She smeared it liberally all over her arms, face and so forth. The trouble was it didn't smell like citronella and when some light was shed on the subject it turned out she had used Gentian Violet! She spent a couple of hours trying to scrub off all the purple and a couple more days walking around in a slightly mauve condition!

Two Close Shaves for me

In the space of three or four days I was nearly killed twice! On the first near-miss, we were pumping out the old shaft when the pump fouled up, so I went down to investigate. The shaft was about 80 feet (24 metres) deep with about 30 feet (10 metres) of water still left in it. As soon as the pump stopped the water started to rise. As I went down into the shaft I could see water squirting out from between the timbers. Unaware of the danger, I dived down about 20 feet (6 metres) beneath the water and discovered where some timber had fouled the suction hose. I cleared that after half a dozen dives and came back out. While down there I noticed all the timbers were really bulging inwards and hadn't realized what a terrific pressure must have been behind them because of the water being sealed up behind the timber. No sooner had I climbed out of the shaft when there was a rumbling sound down below and a great swooshing! The whole mine shaft simply collapsed down around the suction pipe and air driven pump. That shaft was such a mess it had to be abandoned.

We started sinking another shaft in graphitic schist, which is very slippery, like talcum powder. It was very soft but not quite soft enough to dig using a pick and crowbar. It had to be drilled and blasted, but if we placed any heavy shots the graphitic schist would flow in from the sides, just like soup, so we had to obtain some sawn timber to box up the shaft. We ordered 50 tons of 8-inch x 2-inch x 18-foot (20 cm x 5 cm x 5.5 m) hardwood planks from the south. As we went down the shaft we used the planks to force the boxing down to prevent the sides from falling in.

I had sent Frank Fidler, one of my drivers, to bring up a load of timber on the road train. About 5 miles (8 km) from Rum Jungle he was negotiating a fairly sharp bend in the thick scrub. There were gutters in the track and an old fallen tree was buried in the ground from which a sapling had grown out sideways and been chopped off leaving a sharp stake protruding about 6 inches (15 cm). As Frank came around the corner he didn't notice it in the high grass and staked about $1,000 worth tyres within about ten seconds! One line of tyres ran over the stake which blew each tyre in quick succession.

Because he was so late we came back in the car to see where he was. After swapping the spares and putting some more tyres on the road train, we returned into Rum Jungle. Frank, who was a really conscientious man, felt extremely upset for such a thing to have happened. I was trying to pacify him as the road train pulled up near the shaft on a slight slope. Frank was so upset he didn't even want to eat any lunch.

A load of timber bound for 'Rum Jungle' in 1950.
A portion of the front load fell on me while unloading.

We started to unload and had undone the wire ropes which were lying on the ground. Frank was on the other side of the truck and I called out to him, "You pull the rope! I'll thread this side through." As I bent down to pick the bundle of rope up off the ground, about a 3-foot (1 metre) wall of planks (about 18 hundred-weight, or almost a ton) fell on my back and crushed me into the ground, face down into the gravel and black Rum Jungle dirt! (Some of that dirt is still under the skin alongside of my nose). Then a second heap of planks started tumbling down.

When the other men who were unloading the trailers saw what had happened they cleared off around the back of the trailer, afraid to face what they expected to be a heap of pulp oozing out from under the timber. Randle, who'd heard the noise, came running and walked on some of the planks as he tried to see what had happened. I yelled out, "Get off! I'm under here!"

After they removed the timber off me I hobbled up to the mess tent where they cleaned me up to discover the damage. Some bandages were applied and Randle bundled me into the back of the FJ Holden and rushed me to the Darwin Hospital. I don't think the poor old thing had ever travelled so fast because we arrived in Darwin in about one hour! Daphne travelled up in a separate vehicle from Coomalie Creek after she was notified of the accident.

I received deep cuts in the back of my neck and to my face, and also cuts down my back. The tendon on the side of my left knee was also pulled out, which happened while I was doubled up under the wood, with my leg bent sideways. I had actually also sustained a fracture about halfway through my shinbone, though this was only detected about a week afterwards when I continued to suffer pain there. I was very fortunate that a visiting plastic surgeon was in Darwin at the time of the accident. He patched my face up. After I was all stitched up Daphne and I booked into the Darwin Hotel where I spent about a week in bed before heading back to Alice Springs via Hatches Creek.

Dad's Last Joke on the Family

Several months later, after I had recovered and was still working at Rum Jungle, I received the sad news that Dad had died suddenly from a heart attack on 4th April, 1951 while attending the outdoor picture show in Alice Springs with Mum. I flew back for the funeral and returned to work at Rum Jungle.

After the funeral, when all the family and a few friends gathered at home, Mum suggested that Bill Petrick go over to the pub across the street and get a bottle of Scotch whisky so that we could relax and have a bit of a drink to "cheer Dad on his way." Mum and Dad customarily kept a gin bottle full of water in their old kerosene refrigerator for when they wanted some cool water. Unknown to us, Dad had earlier put a real bottle of gin into the fridge.

When Bill returned whisky was poured for those who wanted it and diluted with "water" from the gin bottle from the fridge. I remarked to Randle, "Gee, this is strong!" so I grabbed the gin bottle and watered it down. Some of the others also grabbed the bottle commenting that theirs was a bit strong too, saying to Elsa, who had poured the whisky, "Gee, you gave us a good helping!"

The whisky still tasted far too strong so we checked the gin bottle in the fridge. The water bottle in there was still untouched and we discovered we had been trying to dilute the whisky with straight gin! Mum came into the kitchen and when we told her what had happened, despite her sadness at Dad's passing, she laughed and said, "That's Dad having his last little joke on all of you!"

Mona, Randle, Myrtle, Mother, Elsa, Kurt and Trudy at Dad's funeral in 1951

Finishing at Rum Jungle

After the funeral I returned to Rum Jungle where we were sinking the new shaft using that fateful load of timber which had almost cost me my life. We had gradually worked down to about 80 feet, boxing it as we went, which was a very slow process. By this time the Government had decided they wanted a bigger development there. We could have taken the contract on a ten per cent cost plus basis, but our mining engineer, who was a Yank, fell down on the job. He wouldn't leave Sydney where he had a beautiful blonde, nor look after our finances properly, which he was expected to do. We all decided to let the contract go.

TEP (Territory Enterprises), which was an off-shoot of BHP took it on. During the first week during the handing-over process, they boasted to us, "Well, we've got our mining engineers (experts) here. We'll show you how to put a shaft down at about 10 feet (3 metres) a day." So they sent some of their workers down the shaft and drilled out a full 8-foot cut, putting in full charges of explosives as if they were working with hard rock.

Next morning they sent a bloke down the ladder to start bogging out. He had only descended to about 25 feet (7.5 metres) when he came back up again. They couldn't believe it when he told them the shaft was filled up with stirred up graphite schist from their big charges of explosives. They abandoned that shaft! So their big guns didn't do so well on that one! After trying to dig another shaft without success they finally ended up open-cutting the whole deposit.

I then went over to Hatches Creek to work the wolfram mine there.

Daphne (Johannsen) Little in 1990

Chapter 34
Being 'Conned' by Con

While I was in Darwin recuperating from my accident at Rum Jungle, Pat Davis from Hamilton Downs Station contacted me and asked if I'd call in at Hatches Creek on the way down. Con Perry from Tennant Creek had talked him into taking a half share in a wolfram mine at Hatches Creek. Apparently things weren't going too well and Davis thought that Perry might be "conning" him. Since I had some experience in the mining game Pat asked me to take a look at it and consider taking his share over, paying for it later on if and when things came good.

The track into Hatches Creek was pretty rough, and because I was still quite sore from my injuries, Daphne did all the driving. A fault in the old Holdens, especially for bush driving, was that the plug in the bottom of the petrol tank was easily knocked off, which happened to us. Luckily I heard the knock and we stopped to have a look. I quickly put my finger over the hole to stop the petrol flow, while Daphne cut a green stick to knock into it. We cut the stick off flush so it wouldn't get knocked off again. It was then I first noticed the pain halfway down my leg when I put a strain on it, which turned out to be a fracture from the accident. Anyway, we continued (about 8 miles) into Hatches Creek where I removed the petrol tank and soldered it up.

After looking at the mine I thought it might be a good proposition. At the time wolfram was worth £1 per pound (or £2,000 a ton) which was really big money. I had plenty of mining equipment and so did Con Perry. I also brought up some of my ex-drivers, mechanics and so forth, and after finishing our contract at Rum Jungle, we hopped into this venture. Since a couple of my workers were also mining men I formed a working syndicate and gave them an equal share in the mining instead of wages.

I had another close brush with death at Hatches Creek when I was accidentally knocked into an abandoned, collapsed mine. The old shaft was about 100 feet (30 metres) deep with a funnel-shaped entrance tapering down at about a 45 degree angle into a gaping hole about 25 feet (7.5 metres) wide which was too wide to cover.

We were erecting a head frame for the new shaft we planned to sink near the old one. As we were man-handling part of the head frame off a trailer, it slipped and catapulted me backwards into the edge of this gaping hole. The other workers were greatly shocked to see me disappear down the gaping funnel, to say nothing of how shocked I felt too! However, having always been very nimble on my feet I managed to spin around and run around the tapered edge about 4 feet from the top, running up and out about halfway around the other side! We closed the gate after the horse nearly got out, and erected a fence around the shaft in case somebody else got the bright idea of imitating my stunt!

Another day when I was out bush in the Holden, about 20 miles (32 km) from Hatches Creek, the carburettor started to flood due to a bit of rubbish under the needle valve in the carburettor. I pulled it out and started cleaning it. The flies were really infuriating, crawling into my eyes and nose and as I flipped my hand to brush them away, the needle flew out of the seat and landed somewhere in the spinifex. It was impossible to find so, using my pocket knife, I very carefully cut a new needle from a piece of mulga. That improvised needle worked quite well until I returned to Alice Springs later and installed a new one.

The mining went along quite well at Hatches and when we had extracted about 3 tons of wolfram I loaded it on a truck and took it into town (Alice Springs) to put it on the train. However, as so often happens in the mining game, by the time this first shipment reached Adelaide the bottom had fallen out of the wolfram market, so instead of us receiving just over £6,000 after freight and commission, we only received about £1,200 for the whole lot. This meant we had to abandon our mining operation at Hatches Creek.

When we'd decided to abandon the mining Con suggested I send some of the equipment up to Tennant Creek to his place where he could get a good price for it, which I did. Time went by without me hearing anything further from him. Eventually I went up there to see what had happened to it. All the gear had been sold all right, but his excuse for sending no payment to me was that he "had to give it away" and got "virtually nothing for it", and he "was broke" and all the rest of it.

I made a few discreet inquiries later when I came back again and found out he had actually received quite a good price for the machinery but had persuaded the people who bought it to tell me they'd only paid a small amount for it. I was the loser of about £6,000 there, which I wrote off as a bad exercise since it always meant lost time and money to pursue such matters legally. I decided to let the other workers in our syndicate 'off the hook', because they had come in with nothing and went out with nothing, which was about the fairest deal for them.

Chapter 35
Desert Drama in Lasseter's Country

One morning in 1950, Jim English and Jimmy Prince approached me at the workshop with a proposition. They emphasized the need for secrecy and wanted my assurance that if I didn't want to go along with it then at least I would keep quiet. They would cut me in at one third. Jim English, an old gold mining prospector from Tennant Creek, was to supply the money. Jimmy Prince, a miner and general contractor, knew where the place was. He'd been out there before and nearly perished. I had the equipment consisting of two 4-wheel drive Blitz trucks and an aeroplane. After many discussions, looking at maps and so forth I decided to join them. We were going to have a try at finding gold in Lasseter's country. Our proposed destination was near Mount Buttfield in Western Australia, north of the Rawlinson Ranges. We decided to go out via Mount Liebig because a track went out as far as Mount Liebig Bore at the western end of Mount Liebig Range.

After several weeks of preparation and many discussions working out details, we decided we needed two vehicles. Jim English and Jimmy Prince would take turns driving one truck while Darkie Mansell the Guppa Sultan would drive the other truck. We left a duplicated map at home with my wife Daphne so that if there was any accident and I went missing with the plane, this map would be taken to the Department of Civil Aviation (D.C.A.) in Alice Springs.

Before we left the Alice we had been seen talking together in the pub several times, and the rumour was around seeing us three prospectors together, that we were hatching up some sculduggery. They figured we must be planning some scheme, but they never found out anything else at that time!

Early one morning, late in October, the men in the two trucks set off and reached Mount Liebig that evening where they camped. Next morning, which was cold and frosty, when they went to start up they found that one of the trucks had lost its timing gear. They decided that Prince and Mansell would go to Haasts Bluff Aboriginal Settlement which had a pedal radio. They sent me a telegram reporting that they had broken down and asked if I could bring out a cam gear; they would wait at Haasts Bluff for further word.

Actually, I was scheduled to fly out two days later, but when I received the telegram I put a few things together, picked up the timing gear and took off in the Tiger Moth for Haasts Bluff. After landing there and discussing it with the two men, I decided to fly on to Mount Liebig with the cam gear to repair the truck, leaving instructions with Prince and Mansell that if I wasn't back in two hours for them to assume I had landed safely and come back to meet me at Mount Liebig, which was 80 miles (129 km) back. If I couldn't find a suitable place to land at Mount Liebig I would fly to Haasts Bluff and they would have to take the cam wheel and shaft to Mount Liebig to repair the disabled truck. In those days there were no suitable two-way radios available for fitting into small aircraft, so all instructions had to be carefully thought out and followed. After locating the bore, I circled round a few times and picked out a place in the mulga with a bit of spinifex which looked suitable and landed there safely. That night, just after dark, Prince and Mansell joined us after driving back from Haasts Bluff.

I set to work soon after landing and installed the other cam shaft. We all camped there near Mount Liebig that night. Around 2 a.m. I heard a motor vehicle pull up which I thought might be the bore maintenance crew or someone like that, but when we woke in the morning it turned out to be Billy McCoy, the Superintendent of Aboriginal Affairs from Alice Springs and a policeman. The chap at Haasts Bluff had telegraphed McCoy that Jimmy Prince was out at Mount Liebig Bore on an Aboriginal Reserve. Prince had been out that way by camel a few years earlier without a permit, but they had been unable to catch him that time, so Billy McCoy was after his hide.

I walked up to him and casually said, "Good morning, Billy."

He got a hell of a shock. He hadn't known I was there. Anyhow, it was too late to make any 'beg-your-pardons' so later, when we returned to Alice Springs, we were all to be charged for having been on an Aboriginal Reserve without a permit. Actually, this bore was, according to the map, 3 miles inside the Aboriginal Reserve, but the maps out there had never been properly surveyed and I hoped I could bluff my way through. McCoy and the policeman left.

The men in the two trucks set off towards Mount Lyell Brown, taking about a day and a half to get there. I flew back to the Alice because the others had to clear a strip for me at Mount Lyell Brown (or 'Ilbilla' as it was known on earlier maps). I waited three days to give them time to build a strip before flying out alone from Alice Springs to rejoin them.

At seven o'clock on the morning of Wednesday, October 25th, Jimmy Prince and I took off from Mount Lyell Brown in the Tiger Moth with full tanks, two full jerrycans of extra petrol, a gallon of water and a few days' food, heading out towards Mount Buttfield to survey the area. We were map reading as it wasn't possible to compass read that area out there because the error factor in the maps was anything up to 10 miles out in 30. We therefore flew mainly by noting recognizable hills and reading the maps where particular hills, mountains or lakes were supposedly located.

We flew out to Lake Hopkins, which was marked on the map as a lake, but on the ground it was actually a maze of crusty, salty claypans with sandhills criss-crossing over the entire area. Years ago it had probably been a salt lake but the sandhills had encroached with trees and spinifex growing on them. We assumed that it was Lake Hopkins as it tied in with several mountains we had located on the map. We continued on until we sighted the Musgrave Ranges and Mount Buttfield. We circled around Mount Buttfield to try and pick out a decent claypan or suitable place to land and refuel. This was only our reconnaissance flight and we didn't want to waste any time because we had a limited amount of fuel.

We set our course back over Lake Hopkins where we'd noticed a few claypans on the way out. I thought one of those should make a good landing place, and when we came to the south-eastern side of Lake Hopkins the surface looked quite hard and suitable for landing, with hardly any salt crust. I touched in a few places and found that, on the contrary, it was very soft so I tried the

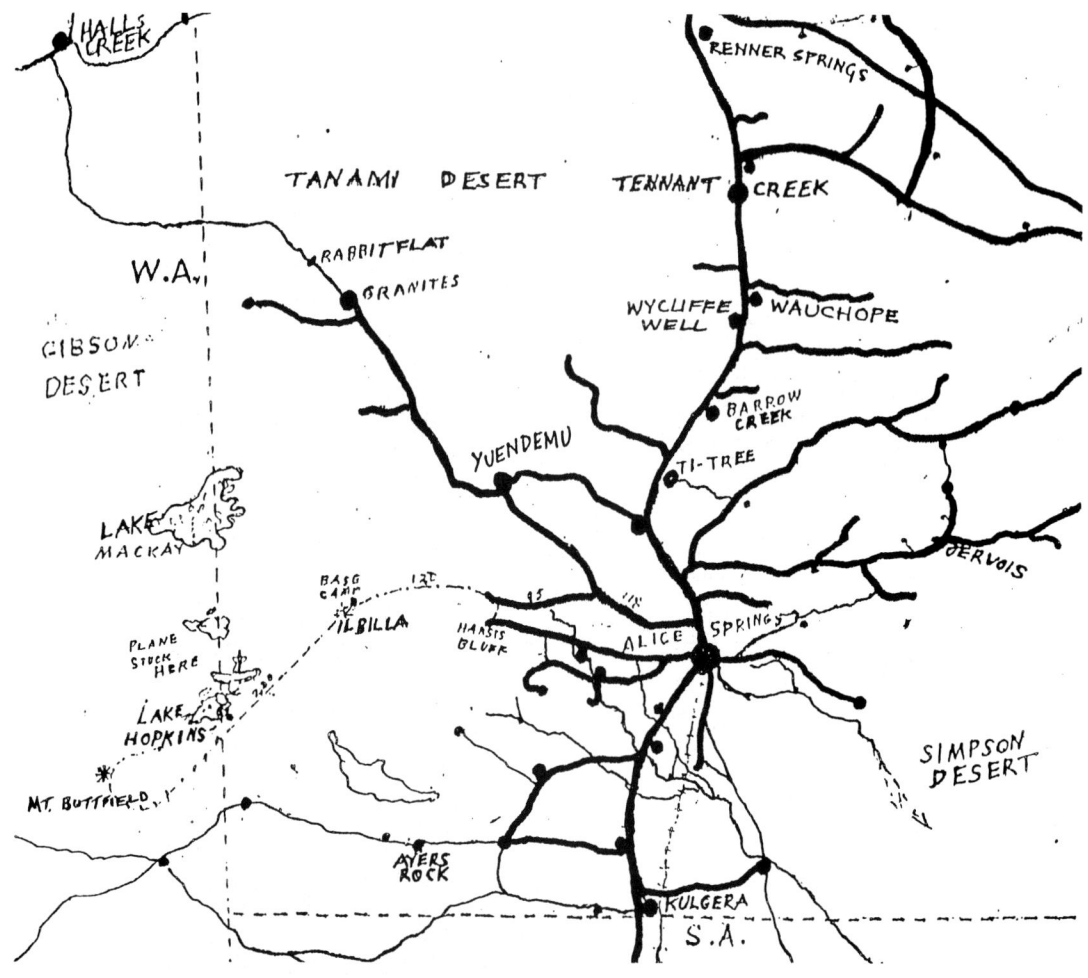

Map of the area where I bogged the Tiger Moth and broke the propeller

sloping side of the lake which seemed quite hard. Around ten o'clock we successfully put down there, re-fuelled and had something to eat.

I then walked up and down the bank and discovered the surface wasn't half as good as it had first appeared. There had been a shower of rain the night before, making several of the little gullies on the side quite soft. I selected an area I thought would give us the maximum run. Jimmy and I climbed into the plane and taxied down to the end. I went to swing it around, but the old Tiger Moth, not having brakes, didn't respond and one wheel dropped into a soft patch. She stuck her nose right into the mud and broke about fifteen inches off each end of the propeller. So there we were, nose in the mud, tail up in the air, without a paddle or any 'Minties' for a moment like this! Jimmy Prince had just been fastening his safety belt when the accident happened. He was thrown forward and bumped his head. He exclaimed, "F--- the bloody aeroplane, the gold and everything else!"

Kurt comparing the damaged propeller of the Tiger Moth with a normal one.

I climbed along the fuselage, grabbed hold of the strut of the tail and swung down on that. My weight pulled the tail back down. I then sat down and thought about our predicament for a while before deciding to take a walk north-east about 6 or 8 miles to where I reckoned the other claypan was. Then, if a search plane came out I could put up markers to show where they could safely land. I walked and walked and walked, but there were no sandhills high enough to provide a view of the countryside. The claypan I'd noticed on the way out must have been much further away than I thought! This proved to be the case when I eventually flew back over the area.

I returned to Jimmy Prince and the plane and decided I would try and find a piece of desert oak with a right-hand twist in it to make a new propeller. I found plenty with a left-hand twist! In any case, it would have been a long, painstaking job chopping a propeller out of a desert oak.

Increasing our water supply was also an immediate concern, so I decided to make a condenser out of two 2-gallon jerrycans to produce drinking water. Using a broken prop end as a shovel, I went out to the lake and dug a hole about three feet deep. First I encountered salt crystals and then the salt brine (saturated salt solution) and baled out a gallon of brine into one jerrycan. I had previously heated the two jerrycans to get rid of all the petrol fumes. I then stood the can containing the salt water in it on the fire and propped the other one up on some sticks with the two spouts facing together, using a little tripod of green stick jammed in to centralise them. In the back of the jerrycan I punched a small hole for the condensate to drip out. As the bottom can began to produce steam, the steam rose into the top container. The breeze blowing on it cooled it and started the condenser going, which produced about six litres of water per day, which was enough for the two of us to survive on and made us both feel better!

I pulled the broken propeller off, balanced it up over a screwdriver and chipped and trimmed away on it with a tomahawk until it balanced. I smoothed the rough edges off with a file and put it back on the Tiger Moth. When I tried it out it seemed to function reasonably well as far as balance was concerned. I started the plane and inched it out of the bog. Of course there was very little grip left in the prop, but nevertheless I prepared for a dawn take-off the next day.

I went for another walk and noticed thousands of birds – budgerigars, waxbills and finches – coming from the north-west in great flocks. At first I thought there must be water in the direction from which they were coming. I'd walked about four miles when I realized there were also other birds such as pigeons and little doves, also heading in the direction of the Petermann Ranges. This indicated that their previous water supply might have been a rock hole which had now dried up. There was no sign of any significant hills – only sandhills and marshy lake areas, so I returned to the plane again.

Early next morning (Thursday) I stripped all the odd bits out of the plane to make it lighter, gave Jimmy most of the food – several tins of meat, tinned peas and a few packets of biscuits – enough to last about a week. I also left him the rifle and bullets plus a couple of rabbits I had shot during my walk on the previous day. I took the gallon of water we'd brought in and left him with the waterbag and, of course, the functioning water condenser.

The air was dead calm on my first try to get aloft – I couldn't even raise the tail of the plane off the ground. I taxied back, clambered out and went for another long walk to see if there was any variation in the terrain around there, but there wasn't any place more suitable for a rescue plane to land. I returned to our camp at about 11 a.m. when a little bit of a breeze was starting to come up. By 11:30 a.m. there was quite a good breeze so I thought I'd give it another go.

We agreed that Jimmy would cut a big pile of green brush to make a smoke signal if he heard the sound of a plane or saw one approaching. Jimmy said if I hadn't returned within a week to ten days he'd cut a good mulga crutch and start walking! I told him to stay near the water condenser and I'd be back.

As I was taxying to take off Jimmy waved me down and did a very wise thing. He came up to the plane, took out his pocket knife and scratched a message in the aluminium of the engine cowling that Jimmy Prince was at a certain point out on the south east corner of the lake. This was a good idea, because if I crashed again and perished any search crews would know where to search for Prince.

The runway I had picked out along the side of the lake was a zig-zag affair. Some gutters couldn't be crossed at the top and some couldn't be crossed at the bottom so I had to memorise the pattern – two at the top, one at the bottom, one at the top, two at the bottom – and zig-zag my way along. There was a fairly sharp turn in the lake bank turning off to the left, leaving half a mile of boggy lake straight ahead. I thought if I couldn't get airborne safely I would put the plane down on the other side, around the corner.

On the last gutter before the turn I didn't cross in the right place and bounced. That bounce tossed the plane up in the air to hang there with the wheels just off the soggy salty ground. I wanted to put it down again but I daren't turn it because I didn't have enough air space. I managed to keep the wheels from touching the boggy lake and gradually increased speed. The rev counter was reading 300 revs which was on the second time around so it was actually 3,300 revs – about 1,000 revs faster than the motor should have been turning! When I reached the end of the lake I had attained a speed of 65 miles (100 km) per hour which was a normal minimum cruising speed so I stuck her nose into the air to rise over the sandhill and the trees on top. I was actually airborne!

As soon as I successfully cleared the sandhill I nosed down again to gather speed, flying just above the mulga trees. I decided I might have to put the plane down somewhere because I feared the motor couldn't stand those excessive revs. Gradually she rose to about 20 or 30 feet above the trees. I kept looking out on both sides to see if I could spot any suitable place to land again, but I didn't dare make a sharp turn. There were several places that might have been suitable but I'd already passed them by the time I could actually see them properly. I had only travelled three or four miles but it seemed as if it had taken about half an hour.

I spotted an eagle ahead, just a bit over to one side. I knew he was soaring on a thermal so I carefully steered under him and picked up a good thermal, giving me a lift of about 80 to 100 feet up where I could see the surrounding countryside a bit better. I decided to circle around and if I couldn't find anywhere to land I would return to Lake Hopkins. I made a gradual turn and started to throttle back to 2,500 revs, but the plane began to sink again. By this time my nerves were really on edge, so I whacked the throttle open again and flew on further, gradually throttling back to about 3,000 revs where the plane seemed to be able to hold its height.

A few miles further on I saw three more eagles wheeling so I flew directly under them and circled. With the lift from that thermal I reached about 200 or 300 feet – it might even have been a bit more. At the time it felt as good as 1,000 feet! I suddenly felt safer and, as my nerves relaxed and the terrible tension eased, tears began to run down my cheeks. I made a decision to continue on toward Mount Lyell Brown.

Up higher in the cooler air I managed to throttle back to 2,700 revs and she held her height there. I picked up a couple more thermals and eventually reached about 1,000 feet above the ground. I passed the big claypan which I had spotted on the way out. It must have been about 15 miles (24 km) away from where we set down. Anyhow, I was quite content to simply head for Mount Lyell Brown which was about 130 miles (210 km) away.

30 miles (48 km) away from the Mount the petrol gauge indicator disappeared which meant the fuel tank was nearly empty. I stayed calm because I knew I had enough height to glide down and be able to pick out a place to land on a spinifex flat without trees. As I flew over the top of the hill I saw the Mount Lyell Brown camp with only one truck there, which indicated the others must have already gone to Haasts Bluff to relay a message that we were missing to the Alice and raise an alarm. I had left instructions with them that if I wasn't back by noon the previous day to raise an alarm.

I circled down, landed safely and taxied back near to the camp. For curiosity sake, after landing, I put a bottle under the drain cock of the fuel tank. A bottle and a half of petrol came from it so I didn't have much fuel up my sleeve! The oil tank was almost empty too. Guppa Sultan was waiting at the base camp and told me that Jim English and Darkie Mansell had gone to send a message. I camped that night with Guppa at Mount Lyell Brown but I must have been completely exhausted as I can't remember anything at all about it.

The truck which was left there had a broken oil pipe. I later found out that when twelve o'clock noon came on the previous day, Jim English and Darkie Mansell decided to wait an extra hour until one o'clock before leaving. They had only travelled about 10 miles (16 km) before they struck trouble. The scrub which they had been pushing down on the way out, was now lying down facing the truck and one of the sticks facing them became hooked up underneath the engine and broke the oil pipe. They patched it up a bit, returned to the base camp and took the other truck to go to Haasts Bluff and have the message sent home that we were missing.

The news that Prince and I had been missing since noon on Wednesday reached Alice Springs by pedal radio on Friday morning around 9:30 a.m. An R.A.A.F. Lincoln bomber, coming from Amberley through to Darwin on a training flight was diverted to Alice Springs to carry out a search. When they flew out to begin searching, what they thought was Mount Lyell Brown was actually another mountain about 60 miles (97 km) west – the maps were so far out! These chaps weren't used to map reading and had only flown using radio compass .

I was still at the Mount Lyell Brown camp on Friday when around dusk I saw the Lincoln bomber fly past about 8 miles (13 km) south, heading west, but they didn't see me or my signals. Guppa and I had put up a big smoke flare by stacking all the bushes which we'd previously stripped off for the landing strip. I poured a couple of gallons of petrol on it, which made a helluva big flare and plenty of smoke when I set it alight. The R.A.A.F. crew said afterwards they had actually seen the smoke but thought it was an aboriginal camp fire. So much for our smoke signal! Anyway, the Lincoln bomber circled right around down to the Petermanns and saw a few more small fires which were aboriginal camps, before returning to the Alice and reporting they hadn't found us.

I blocked the truck's oil pipe off as it only led to the oil gauge. I then set off alone in that truck and drove all night until at about one o'clock Saturday morning when I met Nick Hunter and Carl Koerner in 'Bertha', the Diamond-T prime mover of one of my road trains. They had 10 tons of aviation fuel on board for delivery to Haasts Bluff. There had been some thunderstorms in the area and Nick and Carl were bogged in one of those sloppy places about 30 miles (48 km) before reaching Narwietooma. They had almost extricated 'Bertha' from the bog when I arrived. I put a rope on her and helped them snig it out and everyone turned back to return to the Alice.

Once he'd heard the news that we were missing, Pat Davis in Alice Springs had insisted on joining the search even though he was practically bed-ridden and crippled up with arthritis. The Department of Civil Aviation (D.C.A.) didn't want to let him go because of some high level thunderstorms out that way but Pat told them where to get off! He insisted, "Stop buggering around about weather reports and bloody get going. Look at the weather coming in! My mates are out there – I'm going!"

Pat left Alice springs late on Friday afternoon and flew out to Haasts Bluff to refuel and start searching the next morning. Still wearing his pyjamas he was helped into the Gull by Carl Koerner. He circled Mount Lyell Brown and saw my plane on the ground so he radioed back to Alice Springs to say that my plane was on the ground but he couldn't see anybody near it. He therefore assumed that Jimmy Prince and I had arrived back safely and gone with the other vehicle. As he circled around and did a second swoop to have a look at the cleared strip, he realized it was too short for him to land, so he climbed and headed back to Alice Springs. He'd only flown a few miles on his return journey when the Vega's motor started to play up – it cut out and back-fired. His radio was still switched on when this occurred. Pat was apparently quite fluent with his language which all came over loud and clear on the radio being monitored in Alice Springs! Eventually, by pushing and pulling everything the blockage half-cleared and he gained about three-quarter power, so he didn't touch anything else. He simply climbed and climbed and climbed until he reached about 8,000 feet, arrived over the Alice and landed.

When the news that we were missing reached the D.C.A. on Friday, all the journalists swooped. There was a plane lost, blah-blah-blah (great excitement!) and newspaper reporters came up from Adelaide, plus the local ones. Every newspaper in the country had a field day on the story. Of course that meant that the cat was out of the bag about our destination – it was in Lasseter's country – another plane searching for Lasseter's Gold Reef and so forth! I still have a few of the newspaper cuttings.

When I arrived back in Alice Springs early on Sunday morning, 30th October, I tried to rest after making a few phone calls to try and get another plane or spare prop to return and rescue Jimmy Prince. I eventually arranged with a friend of mine, Ross McKay, who owned another Tiger Moth, to fly me out to Mount Lyell Brown base camp where my plane was. McKay didn't like the idea of going out there at all but eventually I convinced him that I had to go and pick up Jimmy. I borrowed a prop from another plane, stuck that in diagonally in McKay's front cockpit and guided him out to Mount Lyell Brown. He was quite a competent pilot but understandably felt very nervous about going into the inhospitable western desert country. We landed at Mount Lyell Brown where I then flew McKay's Tiger Moth plus the spare prop, out to Lake Hopkins to pick up Jimmy Prince while McKay waited at Mount Lyell Brown for us to return.

Page 8 — THE ADVERTISER, Monday, October 30, 1950

MISSING FLIER AT ALICE

Companion Still At Lake

ALICE SPRINGS, C[...]
Kurt Johannsen, one of tv[...] had been missing in Centr[...] since Wednesday, arrived in [...] today by truck from Mount Ly[...]

Repaired Propeller With Tomahawk

MR. K. JOHANNSEN'S STORY OF FLIGHT

How he re-shaped the smashed ends of the propeller of his 'Tiger Moth' aeroplane with a tomahawk and a hacksaw blade after the plane had nosed over on the edge of Lake Hopkins, and how he limped in to Mount [...]

Prospector Rescued At Desert Lake

ALICE SPRINGS, Oct. 30.
Kurt Johannsen flew to Lake Hopkins today in Mr. Ross McKay's Tiger Moth and picked up James Prince, 70, taking him to Mount Lyell Brown.

THE ADVERTISER, Saturday, October 28, 1950—

Men Still Missing In Search For "Lasseter's Reef"

DARWIN, October 27.
A search by air and land has so far failed to discover any trace of two men who have been missing in a Tiger Moth plane since Wednesday in the waterless Central Australian Desert.

Some of the newspaper headlines about our adventure from The Advertiser and the Centralian Advocate, 1950

This time I knew how treacherous the landing was at Lake Hopkins. D.C.A. hadn't wanted us to go out because of the high-level thunderstorms all around the area. They were right – the storms were building up and coming across from the west.

Before I'd left him out there one of Jimmy's biggest worries had been that I wouldn't be able to find the place again. As I approached the lake, I spotted the place where Jimmy was waiting from about 20 miles (32 km) away. I flew low down and did a circuit at first so as not to excite Jimmy too much and, to show him that I could see him, I waved. He had been lying down in the shade of a bush, but as soon as he heard the plane he hopped up with his 'gammy' leg and hobbled to the fire, pulled out a fire stick and tried to poke it into the spinifex bushes to make some smoke but the fire wouldn't catch. I continued to circle around. By this time a gusty wind was blowing diagonally across the landing strip so I had to come in across the soggy lake and touch down right on the edge and swing it around. As soon as I touched the ground it had the same effect as putting on brakes – the Tiger Moth pulled up abruptly on the spongy ground, within about 50 metres. As I taxied up Jimmy came hobbling towards the plane. As I shut the engine off I could see two clean white patches down his cheeks where he had been wiping the tears of relief away – he was so happy to see me! He told me he had planned to wait another week and if I hadn't returned he had picked out a piece of mulga to fashion a crutch and start walking towards Ilbilla Rockhole. We never did know for sure whether Jimmy actually had a wooden leg or was simply lame. He often referred to his gammy leg as his wooden leg but nobody ever saw it.

The condenser, which he had kept going continuously, was still full of water and running over. I emptied the precious water out of the tins and tossed them in the plane saying, "Come on! We've got to get going quickly because there is a thunderstorm coming." Jimmy was quite upset about me tipping the water out but I still had a gallon in the plane. I helped him in and we took off for an uneventful trip back to the base, other than dodging a few thunderstorms. After refuelling, I fitted the spare prop onto my Tiger Moth and Ross McKay and I set off back towards Narwietooma, both planes flying together. Jim English, Jimmy Prince, Darkie Mansell and Guppa Sultan packed up the camping gear and fuel drums and returned to Alice Springs in the truck. McKay and I landed at Narwietooma, had a cup of tea with Sam Calder and his wife Daphne, before we continued on to the Alice.

The worst part for me had been going back and landing on that treacherous place where Jimmy had been waiting! After I arrived back home I tried to sleep but every time I dropped off to sleep I would wake with terrifying nightmares of all sorts of mishaps. The telephone didn't stop ringing for several days, with newspaper reporters wanting to know all the gory details and many friends also ringing. It was two or three months before the nightmares disappeared.

As a result of all the publicity a group of people from Sydney arrived and wanted to go out there. Billy McCoy from the Department of Aboriginal Affairs, Bob Coxen, the Director of Mines, and some others made up a party. Carl Koerner also went out with them. I didn't want anything more to do with it in such treacherous country.

They all set off in one of my 4-wheel drive trucks and a couple of other vehicles, going out via Ayers Rock. When they arrived at the Petermann Ranges they headed west but became tangled up in the big sandhills going right up the side of mountain ranges, and getting bogged. They became very disheartened and 'tossed in' the expedition. One truck broke a piston from over revving and over-heating. Luckily my truck was carrying a spare. One chap from Sydney had what he called his "gold magnet". He claimed it could locate gold anywhere within 20 feet (6 m) of him. Bob Coxen said he had a piece of gold in his swag to show the aborigines out there what gold looked like, so they'd recognize it if they found any. When this Sydney bloke was asked to demonstrate how he would find gold he never picked up the gold in Coxen's swag!

Eventually Jimmy Prince, Jim English and I were summonsed to appear in Court to answer the charge that we "had been on an Aboriginal Reserve without a permit." To short-circuit the others getting in trouble I became spokesman in the Court and said I would accept full responsibility in taking the charge. I stated that the others were all working for me. I then pointed out that the onus was on McCoy to prove that we had actually been on Native Reserve land. The maps, having never been properly surveyed, were very inaccurate. I showed the Court where Haasts

Bluff Aboriginal settlement was, according to the surveyed map, right on top of a range of hills! McCoy couldn't prove anything within 10 miles (16 km) of where the boundary was marked. The case was dismissed, which was the end of that charge!

A Sequel

Several years later Sonny Woods, a very intelligent aboriginal man who was working for me, married an aboriginal woman who came from the desert country out west. As a young fellow Sonny had belonged to a tribe in that area. However, he had moved into 'civilization' and was an excellent worker.

The old chief and a second chief from his wife's tribe out west had come into Alice Springs, driven in by the drought. They'd first trekked into Haasts Bluff settlement but half their tribe had perished on the way. All the waterholes and soaks had dried up, even the permanent waterholes which had never been known to dry before. They made a dash for the settlement but only the reasonably fit people managed to reach Haasts Bluff.

Anyhow, the old chief and second chief were talking to Sonny Woods, asking him what he did and where he worked and so forth. Sonny was telling them everything, including the fact that I had an aeroplane. They nodded their heads and commented, "Oh, somebody came flying out into our country over Sladen Waters."

Sonny said, "Oh yeah, that was my boss. He was out that way – he went out there looking for gold and broke his aeroplane on the way back."

The old bloke said, "Oh yeah, I know what he is looking for – I know where it is."

Next day Sonny brought the old men around to see me. The chief couldn't speak much English but he was a very intelligent, arrogant old chap with lots of battle scars on his legs, back and chest. After deciding he trusted me, the chief said he would take me out there and show me where the gold was, subject to me helping him get his people back out there again. In return, he wanted a bore, a windmill and a big mob of goats. They cross-examined Sonny Woods for quite a while again. Sonny told me the old chap wanted to make sure I wasn't connected with any "Gov'ment shit", which was what the old chap called people from the Government, Police or Aboriginal Affairs. He wanted nothing to do with them! He reckoned all they did was to make promises and then take their country away and give it to "white fellas".

When he was satisfied that I was 'all right' he had another meeting with Sonny and me. I promised Big Foot 400 goats and the windmill from Jervois if he would show me the place where the gold was. He decided that it would be okay if Sonny took me out there – he would show me where the place was. Big Foot then decided he wanted more proof that I would keep my promise. He wanted to actually see the nanny goats, so we drove out to Jervois for him to see the goats. When he saw them he nodded his head "Yes" – that was very good – he was really happy. Then he suddenly got a 'hang-dog' look on his face.

Sonny said to me, "He is not happy."

I asked, "What is wrong, Big Foot?"

He replied, "Toobar! Toobar!" He meant it was 'too far' to walk the nanny goats the 230 miles (370 km) from Jervois to Alice Springs and then another 300 miles (480 km) out to Docker River. He shook his head and repeated, "No good, no good, no good!"

We tried to explain to him how we could put them on the road train and transport them out there, but he couldn't understand that concept at all. When we returned to the Alice we pulled up at my workshop. Just as we arrived back one of my road trains with a full load of cattle was pulling out to the trucking yards to unload them. When Big Foot saw them a big smile spread over his face as he now understood how we would solve the problem of taking the goats out that far.

I said to him, "We put the nanny goats into that one."

The old fellow nodded and said, "Oh yeah, yeah, yeah! Vellygood, vellygood!"

Because I had been in trouble with the police and McCoy for not having a permit on the previous trip, I went around to McCoy to apply for a permit for Docker River on the Western Australian side. McCoy only had jurisdiction over the Northern Territory but he refused to give me a permit to go on the Aboriginal Reserve in Northern Territory unless he came too and got a cut in whatever we found.

Old Big Foot refused to have anything to do with either a policeman or McCoy coming – only Sonny and me. He didn't trust McCoy or any "Gov'ment shit!" He was also becoming nervous sitting around in the Alice for too long. To keep him happy I drove him, together with Sonny and a couple of their women, about 25 miles (40 km) out of town in the scrub where there was plenty of good hunting for kangaroos. I gave Sonny a rifle and left them there for a couple of days until I could arrange for McCoy to contact the Chief of Aboriginal Affairs in Canberra for a permit. My application to obtain a permit was refused by Canberra. I found out later that McCoy had apparently worded his telegram in such a way that he "wouldn't recommend granting it unless he went out with the party concerned."

Although I was disappointed about that I still decided to help the old chief and his people return to their own county. I went out to the scrub, picked up the aborigines and brought them back into town. I decided to buy some camels, as Sonny said he could take them back out west and try and see if he could find some samples of gold for me. I bought five camels with pack-saddles and gear from an old chap who was advertising them for sale. I loaded them up with about three months of rations and water cans and off they went, happy as Larry. They travelled out via Haasts Bluff and collected the rest of their tribe to go back to Docker River.

On the first night out of Haasts Bluff, Sonny became really nervous because he was in 'hostile' country. He and another boy were so scared they left the camels and all the gear with Big Foot and his tribe. Sonny returned to Haasts Bluff while the rest of the tribe trekked towards Docker River. Of course, none of them had previously had anything to do with camels. About 30 miles (48 km) further on they made camp and didn't hobble the camels because they were afraid to go near them when they made funny noises and so on. All the pack saddles were just left there. They took whatever stores they could carry and set off back to their own country on foot. So that was the end of that venture! Sonny didn't return to town for about a month when he got a lift back and came to tell me what had happened. He was supposed to have brought the camels back to me, but I never saw the camels again. They and their offspring are probably still running around the desert somewhere!

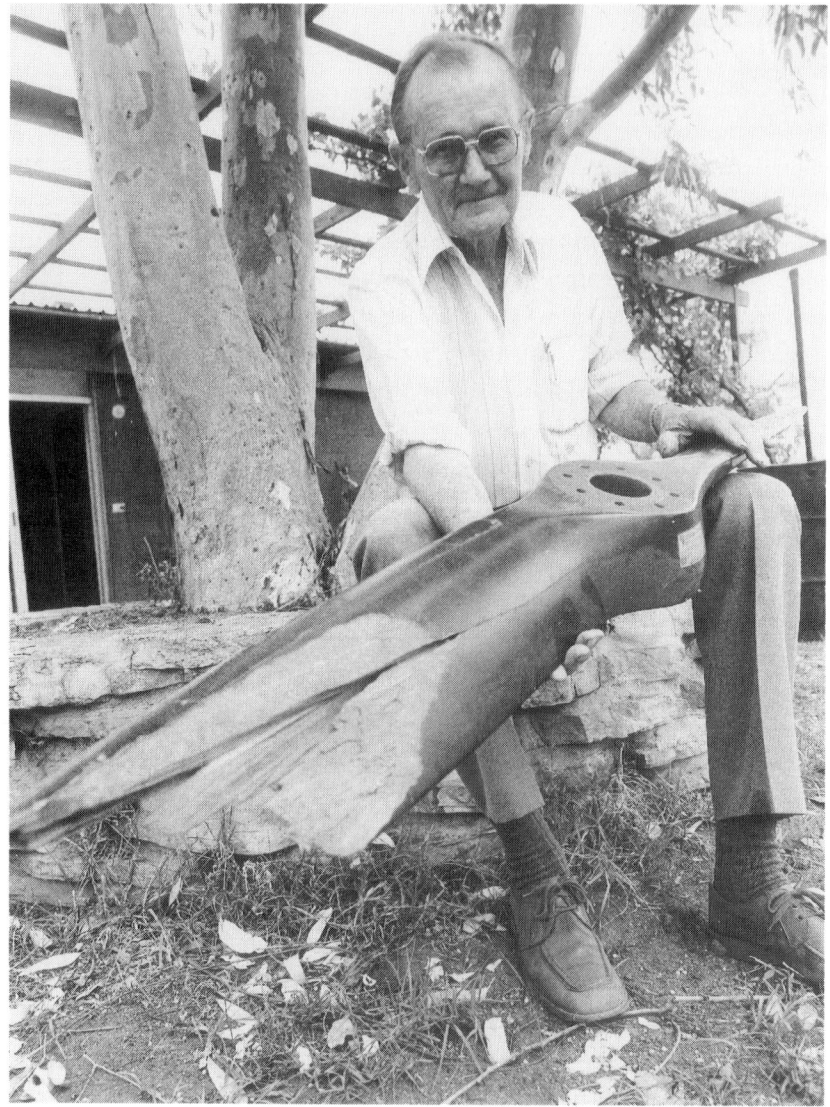

Kurt, 1987, with the broken propellor which is in the Aviation Museum, Alice Springs (Photo courtesy Carmel Sears, 'Centralian Advocate')

Chapter 36
Solar Power

In 1955 I had the bright idea that I could revolutionise electric power generation in a sunny part of the country such as inland Australia or for that matter, any other sunny country. I designed parabolic panels as collectors and experimented with storing the solar energy in the form of heat in storage tanks which contained insulated rock fill. I had the idea of establishing a series of ten solar power plants 400 or 500 kilometres inland, away from the coastal cloud belt, inter-connected to a power grid linking all the cities and towns from say Brisbane to Adelaide. If any power station had a surplus capacity, above its own district's requirements, the load could be shared if some areas were under cloud. Storage of power was designed to last for seven days without sunlight.

The pilot plant I built was too small for economical operation. John ('Black Jack') McEwen, who was deputy-leader of the Country Party from 1943–1958 and later Deputy Prime Minister from 1958–1971, had become quite interested over the years in my varied activities, including building the road trains et cetera, and always looked me up on his visits to the Territory.

When I told him about the solar power plant and showed him my detailed plans he became very interested and said he would introduce me to a thermal power consortium in Sydney, New South Wales. Their engineers were also interested in the concept and feasibility of the plans, but the economists knocked it. They pointed out that, at that time, the cost of fuel (coal) only amounted to about 8 per cent of the total cost of selling and producing electricity. The main costs were in the reticulation and overhead expenses. Of course, the situation could possibly be quite different today with higher fuel costs and the emphasis on reducing the greenhouse effect from coal-fired power stations. My idea probably still has merit.

I was also consulted by Australian Solar Ponds Pty. Ltd. for advice on some aspects of construction of the solar ponds installed in Alice Springs in the early 1980's. They used a salt brine storage area which was constructed and operated by David (Freddo) Frederiksen. He has since constructed two thermal power plants operating from the heat of hot artesian water basins, one at Mulka Station, South Australia on the Birdsville Track and the other supplying electricity to the town of Birdsville.

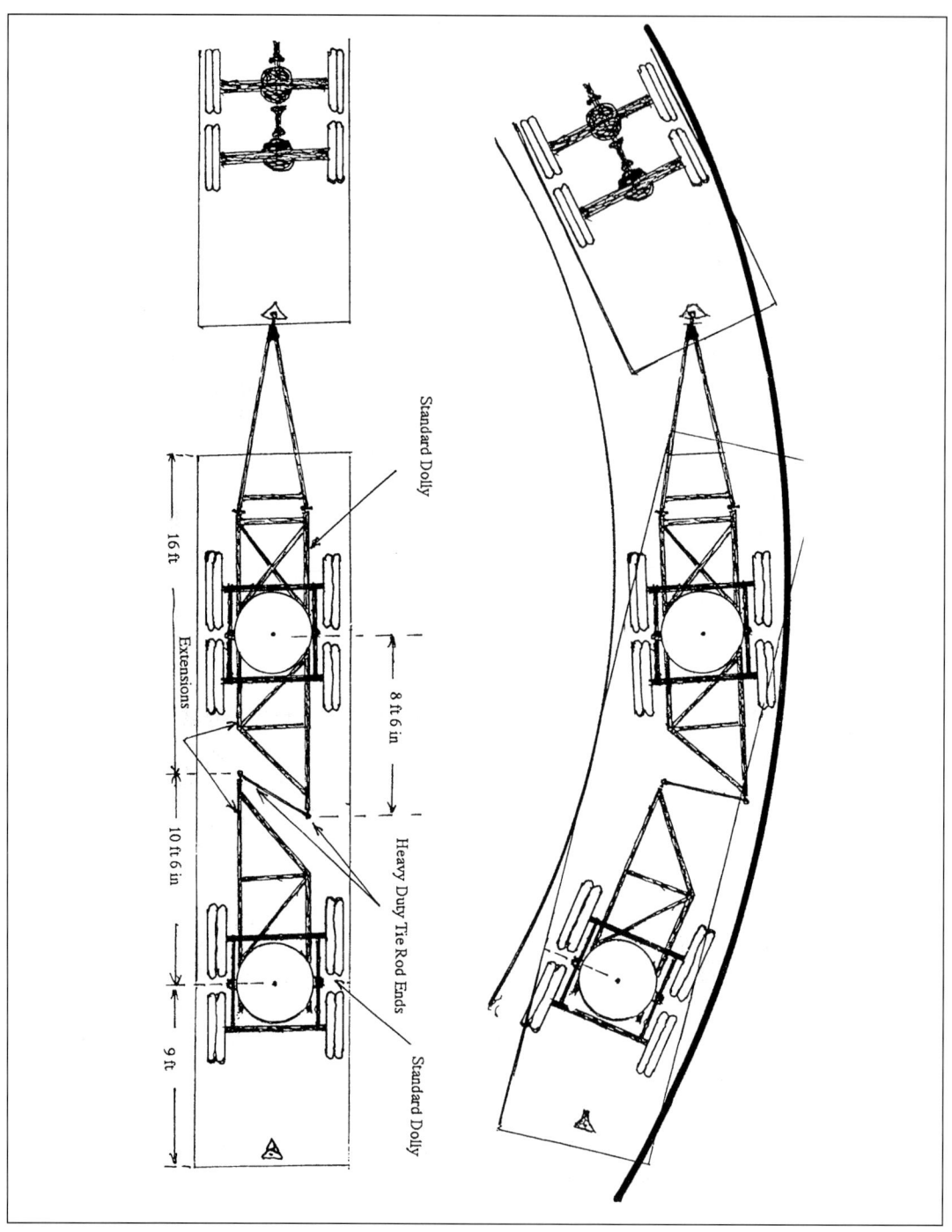

Diagram showing the self-tracking mechanism on my trailers for the first road trains

Mulga Express' Mark IV going up a sand dune in the Simpson Desert, N.T.

Dalhousie homestead ruins, Witjira National Park, South Australia

Finke River Gorge, N.T.

Finke River Gorge, N.T.

The Olgas, N.T.

Rainbow Valley, N.T.

Chambers Pillar, N.T.

Full camping rig of 'Mulga Express' Mark IV

Bungle Bungles, W.A.

Southern Bungles, W.A.

Geikie Gorge, W.A.

Lawn Hills Gorge, Queensland

'Magnetic' ant hills, Litchfield Park, N.T.

Jim Jim Creek escarpment, Kakadu National Park, N.T.

Sunset at Corella Dam, near Mount Isa, Queensland

Deep Well Homestead during the 1920s, painted from memory by Kurt Johannsen, 1987

PART V
MORE TRANSPORT AND MINING

CHAPTER 37
BACK TO CATTLE TRANSPORTING

In 1959, after receiving many phone calls from station owners and agents begging me to put my cattle trains back on the road, I started off again. The price for carting cattle then was roughly double the previous price when I'd quit earlier. Quite a few other transporters entered the market around then because the cattle owners were shifting out thousands of head of their starving stock.

The other truckies and I formed the Cattle Transport Association in 1959, the aim of which was to get uniformity into the prices and service. At first the other truckies didn't want me to be a member because they thought I'd carry too much influence. However, they realized that if I wasn't a member they might not be able to compete with me. I agreed to join and was elected President. I went along with them for several years until one of the bigger transporters decided to go his own way which destroyed the original concept of a fair deal for everyone.

One of the best bookings I ever had for cattle transporting was in the early 1960's when both of my road trains, 'Bertha' and 'Wog', were needed simultaneously for long-distance haulage, with payloads of cattle going both ways – with stud bulls one way and fat cattle the other. Unfortunately, it turned out to be a marathon run for disasters!

'Bertha's' motor was due for an overhaul; I'd already sent away for parts, some of which had arrived, but the crank shaft and the bearings hadn't arrived – they had been over-carried. 'Bertha' badly needed those parts but I had to send the two roadtrains away, 'Bertha' to Camooweal and 'Wog' to Wave Hill (now Kalkarinji, Northern Territory). 'Bertha's' oil pressure was low and she wasn't really fit to take on a long trip.

'Wog' had to pick up a load of bulls from Helen Springs for Wave Hill, then bring a load of fat cattle down from there to Alice Springs, followed by another load of bulls to be carried back from the Alice to Wave Hill and then on to Wyndham in Western Australia. 'Bertha' was bound for Camooweal where she had to pick up cattle bound for Alice Springs, and then pick up another lot from Helen Springs to go back to Wave Hill.

'Bertha' went to Camooweal, where the cattle were loaded. She had only travelled about 5 miles (8 km) when the motor seized up as she climbed a steady grade. The worn parts which I hadn't been able to replace had caused the seizure of the motor. I received a message telling me about that on Saturday afternoon as I was just relaxing after the heavy week's work of getting the trucks ready to send away. I left Alice Springs that evening and by two o'clock in the morning had arrived in Camooweal. I went straight to work on the motor, dismantled it, put the new sleeves in and reassembled it. By nine o'clock that morning we took the truck and trailer out to the yards and started re-loading the cattle.

Some of my road train crew – Bert Noske, Henry Kunoth and Roger Wait, about 1950

As we were loading the cattle a message arrived from Wyndham saying that 'Wog' had broken a tail shaft. I immediately set off from Camooweal on up to Wyndham, travelling all night and arriving there early the following morning. By that time I was feeling pretty tired, to say the least! I mended the tail shaft and sent my drivers on their way again, headed towards Wave Hill.

I wearily set off to return to the Alice, driving over some very rough patches of road. Rocks on the road knocked the U-bolts off my Holden's rear springs, so I had to wire them up. Eventually, when I reached Newcastle Waters, I bought and installed a couple of new U-bolts before driving through to the Alice, arriving early the next morning – the third morning since first leaving Alice Springs on the Saturday afternoon! By that time the spare parts needed for 'Bertha' had arrived so I packed up and headed out to Camooweal, pulled the motor out of 'Bertha', and fitted in the new crankshaft, bearings, pistons and sleeves and reassembled it.

Meanwhile, Lindsay, who had been on a holiday, trying his wings out around the eastern states, arrived at Camooweal on his way home, which was a godsend. He helped me put the motor back and drove 'Bertha' from there to Wave Hill for another load of cattle. As I was getting Lindsay ready to go we received yet another message from Peter Ritchie in Wyndham. 'Wog', which was operating out there, had rolled the 51-foot trailer at the jump-up on the Gibb River road.

I immediately set off back to Wyndham again, which would make two visits to Wyndham inside of a week when I'd never ever been there even once before that week! I arrived the following morning, beginning to feel as if I was on some kind of endurance test!

Peter Ritchie, my head driver up there, had been out in the hilly country driving 'Wog', bringing in another load of cattle. It was pretty rough going and before he settled down to have a sleep he instructed his young offsider, "Now, when you get to the jump-up, pull up and I'll take it down the hill." It was quite a long, steep hill going down to the plain country, zig-zagging through the scrub, and not a very good track. Anyhow, when the young chap got there he thought to himself, "Peter's still having a sleep, so I'll show him how good I am. I can drive this truck too!" So he set off down the hill, starting off in too high a gear. Next thing the brakes became too hot and before he knew it he was careering along at about 40 or 50 miles (65-80 km) an hour. Near the bottom of one of the fairly sharp bends, the 51-foot trailer rolled and tipped out all the cattle and dragged along for a way. Other than losing the cattle, things weren't too bad. It was still on the property of the owners, so they weren't really "lost", but they were certainly gone from the trailer!

Peter went back the next day with the two trucks, put the trailer back on its wheels again, towed it into Wyndham and straightened up a few bits of angle iron, put a couple of sheets of iron on and left again that night to keep up the schedule of the cattle coming in. That shows the strength of the bridge (truss-type) sides I had on those trailers.

I headed off for the return journey to the Alice, finally arriving there the following day, when I slept solidly for nearly twenty-four hours! Altogether, in nine and a half days and nights I had covered over 7,000 miles (11,200 km) besides working most of the daytime on repairs and travelling half the nights. That was the longest mileage I had ever done in such a short time, and during those nine days I don't think I had more than about thirty hours of sleep! In those days I was a tiger for punishment and enjoyed a challenge. Even now I don't think I've changed too much!

Tricky Truckies

Peter Ritchie, one of my drivers already mentioned above, was a blond-haired, happy-faced, clean-cut young man. Another driver was 'Bob-the-Dog' of short, nuggetty stature, middle-aged and a bit inclined to boozing. Both of them were heading for Darwin with two road trains. Peter had an off-sider but Bob was driving alone. They had camped for the night at Renner Springs where Bob had drunk a little too much liquor for his own good.

Next morning they headed off together, with Bob sporting a bit of a hang-over. About 20 miles (32 km) further on they came to some hilly country. One particular hill was known as 'Gearbox Hill' where many a gearbox on an over-loaded truck had come to grief. Peter and his off-sider were pretty well ahead of Bob. Peter pulled up just over Gearbox Hill and sent his off-sider on, telling him he would catch up with him later on in the day. Taking his pogo stick, which at the time were all the rage, Peter watched and waited just over the crest of Gearbox Hill for Bob to slowly grind his way up. As Bob's truck appeared over the crest of the hill, Peter hopped onto his pogo stick and bounced along the road towards Bob, hopping along merrily. He met Bob, but without even glancing at him he just continued hop-hopping along the road and disappeared out of sight down the other side of the hill.

When Bob got over the crest of the hill he pulled up, gave himself a good shake and stuck his head under the water tap beneath the truck to try and rid himself of "seeing things." He wasn't quite sure whether he *was* seeing things or not, but decided he'd carry on. In the meantime, Peter hitched a ride from a passing motorist and told them of his prank. Eventually they passed Bob's truck and caught up with Peter's own truck. Later on in the day both trucks stopped alongside the road. Peter and his mate didn't let on about the prank and waited for Bob to say something. Eventually, Bob casually asked if they had seen or passed a bloke hopping along the road on a pogo stick. They answered, 'No!" They didn't let him in on the joke until about a week later. It certainly helped Bob stay off the grog for a bit!

Another Trucky Prank

Another driver had smashed and rolled a couple of trailers on a road up north, making a real mess of things. A sign was quickly erected near the site with these words, or something similar: "John Ryan's Wrecking Yard." It was suspected that the fellow who put the sign up was one of Baldock's drivers, a very conscientious driver with a fairly good opinion of his own driving capabilities and inclined to be critical and sling off at other drivers who weren't doing the right thing, or losing their loading, or not servicing their trucks properly and so forth.

Some time later on a trip this particular fellow was carrying a big load of sawn timber (oregon planks and general building timber). Up near Stirling Hills he rounded a corner which had a bit of rough patch. Before he knew what was happening some of his load shifted and he lost the entire load off one of his trailers and completely blocked the road.

A few of the other drivers who had previously been chided by this chap, put their heads together and made up another big sign-board about 3 x 2 metres and using very flash writing painted it to say: "Baldock's Timber Yard – No Order Too Large or Small!"

After all the debris had been cleared away this sign was hastily erected at the accident site, facing the bend. Next time the once-proud driver came around the bend, there was the sign! Some people can't take a joke and apparently this chap went crook at anybody he thought might have been involved in the prank.

Some Risks and Dangers

A near-accident happened on my first trip up to Wyndham in the road train for cattle transport. We were heading towards Jasper Gorge on a new graded track. It was night and the road train was sailing along fairly well with Henry Kunoth driving while I was having a sleep in the back. Next thing I felt a few jolts and heard a lot of noise followed by dead silence. I got up to see what the trouble was. Henry had been coming off one of those long, black-soil plains when he suddenly came upon a steep downward dip. It looked to him as if there was a creek at the bottom, but in the glare of the headlights the road seemed to continue out on the other side in a straight line. However, when he plunged almost into the creek he could see a great bank of earth which had been left by a grader, right in the middle of the creek. The track actually turned sharply to the left and followed the creek up, emerging further upstream. The road gang hadn't yet completed that particular section.

Poor Henry had no hope of stopping! He finished up careering down the creek, completely missing the second bend of the 'S'. Before he could pull up he was about 50 metres along the creek 'without a paddle' in a dead end. It took us about four or five hours to get out of that mess. We couldn't reverse with a string of trailers behind, so we had to unhook the prime mover. There was very little room to spare – the space was only as wide as the truck and the banks were about 10 feet (3 m) deep. We chopped a few limbs away and a bit further down we managed to drive the truck up a side of the bank which was less steep than the rest. We then had to get the prime mover around behind the trailers and one by one we hooked the prime mover onto the front of them again.

Approaching the Katherine River Low-level narrow bridge with a load of drums

Another narrow miss with disaster occurred in earlier years when I was coming down from Darwin with a load of about 1,100 drums on. I felt a bit dubious about the Low-level Katherine Bridge which is only a single-width bridge. Coming down the northern approach there are a couple of sharp bends and it is the same on the way up on the south side – it is completely blind until you get onto the bridge itself.

On the spur of the moment I decided to pull up halfway down on the north side, thinking to myself, "Crikey! If somebody is coming from the other way they've got no hope of stopping or passing." So I pulled up there and sent one of the lads across the other side to warn or hold anybody who might be coming from the other direction before we entered the bridge to prevent a crash in the middle. The lad was about halfway across the bridge when a semi-trailer, doing about 50 miles (80 km) an hour came rumbling around the bend and went straight across. Even though he'd been waved down he didn't have enough braking power to stop. The upward bank on the north side slowed him down enough to enable him to pull up within about 6 feet (2 metres) from my bumper! If we had continued straight on through we would certainly have crashed right in the middle and there would have been a helluva mess! Hair-raising experiences like that made the game quite exciting at times! Of course, it was also quite dangerous and potentially disastrous.

A few years after the Cattle Transport Association had been formed Dick Rogers, R.P.M. Transport, (later Tanami Transport), pulled out of the Transport Association. He went on a trip in the off-season, visiting the largest customers. He told them that he had been doing the main part of the carrying and if they didn't give him a contract for carting all their cattle they wouldn't be able to get them shifted during the next season. That signalled the end of the Cattle Transport Association. Everybody pulled out and each transporter went his own way. It then became a real 'rat-race' with price cutting and so forth, which is when I pulled out and retired from cattle transporting. At that time southern carriers were also facing very lean times. Carriers who brought up a load of goods to the Alice or Darwin saw they could get quite a bit of work there. They cut the prices yet again to obtain loads and meet the premiums on their trucks. In actuality they were only borrowing money on their own trucks, many of them going broke in the process, while ruining it for the locals!

Breaking with old traditions is very hard for some people to do. It took about fifteen years, a lot of expense and hard work to convince the cattle men that road transport was economical and also enabled them to send young, fat cattle away in small lots as they became ready to market. From 1947 to about 1960 a lot of the cattle men 'sat on the fence' watching what was happening. Severe drought finally forced more pastoralists to use road transport, which had already proved its value. Beef roads started to appear, which had previously been advocated for several years by Lionel Rose. In the next decade to 1970, hundreds of miles of beef roads appeared throughout the Northern Territory and millions of dollars were spent on large transport companies. By then about 80% of all cattle were being road transported. By 1990 about 99% of all cattle in the Northern Territory properties were served by road trains.

My road trains have become a part of the history of the Territory. I can justifiably claim to have been the builder and operator of the first road trains for cattle transport in Australia. Others followed as the demand grew and the viability of this type of transport was proved in remote areas where there were no railways and long, wearisome distances to be covered.

Nowadays, with good roads, huge double-decker units carrying up to 150 head of cattle and travelling up to 100 km per hour, rumble along the highways. Despite the criticism and ridicule in 1946 from some of the Government experts and cattle men in those early days, telling me I was a dreamer and that it would "never take on other than for stud cattle", they were proven wrong. My dream did come true!

Chapter 38
Mining at Jervois Range

The Jervois Range, which runs for over 50 kilometres, is about 350 kilometres by road, north-east of Alice Springs. It is on the northern edge of the Simpson Desert where extremes of temperature are common. In summer the temperatures can hover around 40 to 45 degrees Celsius for long periods, while in winter it can be extremely cold at night, with sub-zero temperatures.

About 1950 Jim Coppock and his son Lance were mining copper out at Jervois. They wanted me to cart out an estimated 50 tons of copper carbonate ore for the Adelaide Chemical Fertilizer Company. Copper carbonate was used in treating soil deficiencies in Western Australia and in the south-east of South Australia. I took a trip out in the utility via the old road which went through Mud Tank. At that time there was a fairly new track out there passing through Mount Riddock, Red Tank and across to the Oorabra Rockhole, Jinka Springs and then round the back track into Jervois. It was an extremely rugged track and almost impassable for a transport vehicle.

I decided to take on the job of carting the ore, but first we needed to cut a better track through from Huckitta, which was then called Red Tank. Our proposed track would follow through on the south side of the hills along the edge of the desert and into Jervois from the south. It turned out to be a lot harder than I anticipated. There were four wide, sandy rivers to cross which needed banks to be cut down and there were many more small creeks stemming from the Ranges and running down into the desert. In hindsight, if we had gone further out into the desert we would have avoided most of the creeks.

On the first trip out with 'Bertha' it took two days to travel from the Plenty River to Jervois (60 miles). We camped at Oorabra Creek on the way out and when we climbed out of our swags the next morning everything was covered with frost. I estimated that the temperature must have been about minus 15 degrees Celsius. We had to light a fire under the engine of the truck, stoking it up for about two hours before we could thaw it out, which made us late arriving at Jervois. Because we were so late, we loaded the truck with the copper carbonate as soon as we arrived in the evening, working by bright moonlight. We left at about one o'clock in the morning and at sunrise we had only returned as far as Oorabra, our campsite of the previous night, after about 40 miles (64 km) of very slow driving. We noticed that some of the gum trees and whitewood had been killed by the heavy frost of the night before.

That track took several years to beat down. After some rain it started to form a more solid surface, but was still a very rough track. After traversing the track three or four times we could complete the trip in about three or four days. These days, with the new road, the trip out and back takes only a day.

A Brief Early History of Jervois

About 25 years earlier, in 1928 or 1929, Tom Hanlon and his mate had come across from Queensland, cattle duffing (mustering stray cattle) in the back-blocks around Jinka Springs and

Picton Springs. They headed off around the back of the Jervois Range to Unka Creek which passes through the present Jervois Copper Mine area. They reached a fairly big waterhole there where they spelled their horses, camped and settled the cattle down for a few days. During that time the horse tailer found some samples of copper ore and took them back to show Hanlon, knowing he was an old mining man. That was the beginning of the Jervois Mines.

The track was pioneered by some of the first motor vehicles in the 'Rush' which followed the discovery of Jervois. The original track, which Hanlon used, was only a horse pad from Tobermory Station on the Queensland border. There were a few relics from the 1928-29 Rush left near the track and at Jervois. One was an old Vulcan car parked right in the middle of the Arthur River, where floods had gone over it. It must have become stuck in the river by someone travelling over from Queensland and had been abandoned there. Some 20 miles further on, up in the limestone hills, we came across an old burnt-out Reo making us conjecture about from where and when it was travelling and how it may have broken down. Or perhaps bushfires burnt it out later.

There were some big-hearted, courageous people in those early days, coming right through from Queensland, not knowing where they'd find water or food supplies. They just followed a horse pad using their crude maps to guide them, hoping for the best. Most of them got through, but there must have been many a long trek on foot and occasionally people perishing along the way.

Going up a steep grade with a load of copper ore

The Start of my Mining at Jervois

Some leases at Jervois became vacant in about 1952. Northern Drillers also abandoned their lease which I later took up in my name. There were also a couple of other abandoned leases which I pegged and gradually, over a period of time, bought up the rest of the leases. The minerals contained in the leases at Jervois Range included copper, silver-lead and scheelite, but as there was no profitable market for scheelite and silver-lead after the war I concentrated on the copper.

I had left Bert and Helen Young working a copper concentration plant out at the Pinnacles in the Strangways Ranges, but it was too much for Bert to cope with alone. We closed that operation down and I moved most of the equipment to Jervois where there was far greater potential, although it was quite a battle getting started. It was during the heat of mid-summer just after some big bushfires, so everything was blackened.

A couple of weeks earlier, before we actually moved out to live at Jervois, I went out to put a bore down. We used the road train with two trailers carrying all the mining gear from the Pinnacles, plus 200 sheets of corrugated iron, a pump jack and engine, plus a water tank. On the way out we bought 200 goats from an Italian mica miner in the Harts Range which we used to provide ourselves with fresh milk and meat.

After we loaded the goats up from near Mount Riddock it rained like hell, which was really a godsend because the bushfires which had swept through the country previously had burned out everything. However, it slowed our travelling down, and even though it was only a little over a 100 miles (160 km) from Mount Riddock to Jervois, it took us nearly three days to get there, moving tortuously from one bog to another. In one place we became bogged so deeply on one side with the trailer that the goats stuck their heads out through the crate boards and ate the grass and bushes on the side of the road! Along the way out we cut bushes and threw them in to feed the goats and used buckets to provide them with water. We also milked the goats along the way. Those goats thrived out there and bred up to a herd of 500 in four years.

Our first priority was to get the pump working and put the water tank up on some 44-gallon drums, stacked three high, which we accomplished by using jacks. We also erected a corrugated iron shed with spinifex on the roof to act as insulation to help keep it cool inside. Within a week we had a shower room under the water tank and had started a garden, fenced in with corrugated iron sheets

Initially there were ten of us out there – my wife Daphne, my son Lindsay and me, Mark Ryan, two aboriginal men plus their wives and two children. Daphne drove the old Holden out while I drove the road train. We brought out a kerosene fridge which we used for keeping meat and perishables and settled in. But as soon as the ground dried a 'black' dust storm occurred, blowing up all the soot and ashes from the earlier grass fires.

The second week we built a baking oven out of rocks and clay and the men commenced digging the copper ore to produce copper carbonate for the fertilizer industry. It all had to be hand-sorted and drummed for later transporting to Adelaide and to Western Australia.

Another task was clearing an airstrip. Connair had started a station run and planned to include us for a weekly mail run. However, we had to have a radio transceiver before they'd land. We put the strip in anyway, clearing timber, just in case of an emergency. It was about two years before we could afford to buy a transceiver and be included in the mail service. By that time we were well established and mining quite good quantities of copper ore. We also put in a small earth dam as a rain water catchment. The bore I'd put down to 230 feet was good water for stock but was fairly hard for domestic use.

Soon after we started at Jervois Jack and Edna Maskell joined us to work there and thoroughly enjoyed it. Jack, who was an asthmatic from Melbourne, had first come up to the Territory to work for Connair. He'd applied for the job as a little side bet one evening when they were having a few beers and a game of cards. At the time Jack couldn't even run 50 yards after a tram without collapsing from asthma. Anyway, he got the job and they decided, "Let's have a go!" They received a free fare up and if they didn't like it or the company wasn't satisfied they'd only have to pay their own way back. Jack worked his required twelve months for Connair as a specialist welder and from the day he arrived up there he never had another asthma attack.

When they came out to work at Jervois it was the roughest, toughest work Jack had ever done and also living conditions were quite primitive, but they were both really happy. Later on, when I had to close down because of the drop in copper prices, Jack worked for me in town for about twelve months, building up a crushing plant. He eventually established his own welding workshop in town.

In between a lull in mining activities when prices were low, we left Jervois and went back to town where I carried on with general carting and transport contracts. After about eighteen months copper prices rose again so we moved back out to Jervois. We better established ourselves by building three bush timber and iron shacks. The garden and everything was dead by then so we had to replant it and put down some more bores to try and establish a better water supply.

I'd left an eccentric old pensioner, Jack Stribble, out there to look after things while we were back in town. He was an alcoholic and had wanted to go out bush where he couldn't have access to the grog. He replaced some of the timber in the shaft at 'Hanlon's Reward' mine but he'd become bored out there alone with little to do, and had a bright idea. He decided to start the air compressor going and make an adit at the bottom of the hill going to nowhere! I'd left about a ton or more of gelignite in the magazine and when I returned after about six months to see how he was going, he'd made the tunnel about 20 metres deep, using up half the gelignite, run the air compressor out of order and also cracked the cylinder head. When I asked him why he'd put the tunnel in he said, "I thought I might find some more copper under the hill!"

Cutting a Road to Mount Isa

When the grade of copper carbonate which we were supplying to South Australia for fertilizer diminished and the fertilizer company demanded a higher grade, we couldn't supply them, so I decided to open up a road to Mount Isa Mines which could use our ore as a fluxing ore since it had a high silica content.

Setting off with two trucks, eight men, picks and shovels, axes and about two weeks' rations, we started clearing the track which had only been a jeep track part of the way. We created some new sections, cutting down some of the creek banks. Some of those banks had previously been almost straight up and down and after about ten days of hard work we reached Mount Isa.

While we were cutting the road a fire started under the truck due to grass seeds getting in around the exhaust. I quickly put the fire out by smothering it with sand. Once the road was cut we carted our ore into Mount Isa. That season we carted about 2,000 tons across with each vehicle averaging about 50 tons a trip. We used to go through Manners Creek and Urandangi, a distance of about 310 miles (515 km) each way. The roads were full of bulldust and very chopped up but as Mount Isa wanted as much as we could produce we carried on until the Wet season started. They said that if we could bring two or three times as much the next year they would be very happy to take it.

In the meantime I had spent about £10,000 on doing up parts of the road, putting two new motors in the trucks, purchasing new tyres and so forth. After the Wet ended in May we loaded up the road train and drove over to Mount Isa with our first shipment of 90 tons for that season, anticipating a good financial return.

When we arrived the people at Mount Isa Mines asked, "Didn't you get our letter?" and added, "We have put you on a quota now because we have found a big supply of the same sort of fluxing ore as yours on our own leases, just under the office block at Mount Isa itself. You will only be allowed to bring in 30 tons a month."

Since I had already delivered 90 tons on this first load they said, "We'll accept this lot, but we'll put it in a stockpile and that'll be your quota for the next three months." It was quite out of the question to continue on that basis because it ruined our whole budget and was completely non-viable.

90 tons of copper ore from Jervois, bound for Mount Isa, around 1960

Copper Sulphate Crystals

I decided I'd have to do something else, so I started experiments with extracting copper out of the ore. Copper sulphate prices were very good at the time so I started making little pilot tests and found that I could leach the copper out with sulphuric acid, evaporate the water off and obtain copper sulphate crystals. The evaporation rate at Jervois was very high – around 13 feet in twelve months – an average of about half an inch a day. I set up a leaching plant, a small one at first, using acid to dissolve the copper out of the ore and sold the copper sulphate to the fertilizer industry. That plant worked quite well, so I set up bigger evaporating bays, a little crushing plant and established some vats. That also worked quite nicely and I could produce about a ton a week. At the time copper sulphate was worth about £900 a ton which was a good price and we could make ends meet on that amount.

A Lost Petrol Tank!

One summer, just before we closed down at Jervois Mine for the Christmas period, Mrs James from Tarlton Downs, our neighbouring station, set off into town to pick up her children who were returning home for the school holidays. On her way back home, about 10 miles (16 km) from Jervois, her car suddenly stopped for no apparent reason. She got out to check things out, but couldn't find anything wrong with it. She sent her eldest son, aged about 18, towards our place to get help. About 3 miles away from our place he arrived at the campsite of a couple who were staying on one of our leases – the 'Bell Bird'. They ran him back in their car to see whether they could help or pick up Mrs James and the other children.

In the meantime, while they were waiting for help, the younger children went for a walk back along the road towards Alice Springs, just for something to do. About half a mile back on the track they found a petrol tank half full of petrol, right in the middle of road, just sitting there! They knew they hadn't seen it when they first came along the track so they hurried back to their Mother and said, "Hey, Mum! There is a petrol tank in the middle of the road, half full of petrol. I wonder who it belongs to?"

It suddenly dawned on Mrs James the reason why her car had stopped! She looked under the back end of the car and sure enough, the fuel tank was gone! She had bottomed out on a rough patch on the road causing the suspension strap which held the petrol tank in place to flop off. The tank fell out and skidded along the track without even tipping up or losing a drop of petrol! She didn't know how to attach it back on so she waited for help to arrive. When the chap from the 'Bellbird' turned up in his tip truck he used a steel cable, hooked it on the front of her car and off they went. It is a wonder they didn't pull the car in halves because Mrs James had never been towed before. The cable broke three times – each time the cable took up the slack they nearly broke their necks from whiplash! Eventually they arrived at our place where we put the tank back in place for her. The family stayed the night with us and cleaned up from all the dust they'd collected from being towed.

Always Be Prepared!

A couple of weeks later, when we were all heading into the Alice for Christmas at closing down time, the same couple out there on the lease said they didn't want to come into town with us. They wanted to save money and said they had a bottle of brandy and a couple of flagons of wine. They'd celebrate Christmas out there.

We had only been gone about three days when they changed their minds and decided they wanted to go into town after all. They set off in their little old Fiat, which wasn't really roadworthy for those rough bush roads – it definitely needed another car to be following along behind. They carried no supplies with them except a flagon of wine and a lemonade bottle full of water. They had travelled about 30 miles when the radiator was punctured and started losing water. They turned around and headed back to the lease, after pouring the bottle of water into the radiator.

About halfway back home they were forced to stop again because of the motor overheating. They decided to walk the remaining distance back to the 'Bell Bird'. By the time they were 5 miles (8 km) from their destination the woman was starting to lose her mind from thirst and heat exhaustion. Her husband continued on alone walking back to the camp to get water. Fortunately, Sandy Anderson, who was travelling from the opposite direction, found the woman who had apparently stripped off by this time and was walking around in circles. Sandy took the couple into Alice Springs to the hospital for treatment. They survived, but only just.

We thought the couple had seemed sensible 'Bush' types. It is a foolish and dangerous thing to set off in that kind of country without proper supplies in case of an emergency. Disaster can strike so quickly when people go out unprepared into that harsh climate. Several weeks later, when I went out with the road train to bring back my gear for closing down at Jervois, I also loaded up all their possessions, including their car, and brought them back to the Alice.

Being Blitzed by Mr Blitz

In the middle to late 1950's while still mining at Jervois I learned the hazard of dealing with smaller, less-reputable companies who made promises of lower commission charges and higher profits to the miners. Mr Blitz, a Jewish ore buyer from Melbourne, came up to the Territory to establish some customers. I think he gained some introductions to prospective customers from the Mines Department.

As a result he arranged with two other miners and me to ship our ore to him, which we did in due course. I sent 50 tons containing about 12% copper and 8% bismuth which was worth around £10,000 less freight and commission. The other miners sent him around 25 tons of copper ore, worth about £2,500.

When we finally went looking for our payment, we discovered that Blitz had moved his office from Melbourne to Sydney. After extensive inquiries from all of us we found out that he had gone into voluntary liquidation with no trace found of the ore or him or our money! Before he left Alice Springs he had mentioned he was going to change his name from Blitz Trading to some other name because he thought that many people had prejudices against (he grabbed his long nose) people of Jewish ancestry. The other two prospectors could far less afford the loss than I, although it was a very large loss for all of us. The other two miners had virtually hocked themselves to the hilt and were deeply in debt. It is difficult for honest, hard-working people, who are accustomed to their word and a handshake being as good as a legal document, to recognize the crooks and shysters in business dealings.

Drought Damage

When a drought started around the middle 1950's, Webb Brothers from Mount Riddock were short of feed and water on their station. I suggested to them, "Why don't you bring your cattle out to Jervois, put in a trough and tank and use my bore?"

This they did, fully expecting that in six months the rainy season would arrive and they could take the cattle away. It turned out they stayed there for nearly three years! During this time I was setting up the copper sulphate plant. All around the plant the whole countryside was gradually eaten out by the cattle, rabbits and wildlife, which turned it into a dust bowl. We'd also had another bushfire which burnt the countryside out even further and without any rain a lot of drift sand came in.

When the bigger leaching plant was brought into production the drought had been going on for a few years. Once a week a dust storm would blow in – the desert sand kept drifting in and every time I had a batch of crystals just about ready to harvest a dust storm would blow up during the night and next morning the crystals in the crystallizing bays were covered by half an inch of fine red desert sand. We had to abandon that project for about twelve months in 1958 because everything was being buried in little sand dunes. I closed down, went back into Alice Springs and resumed carting cattle, as described in the previous chapter.

The Hat in the Tank!

Mick Brown was an old-timer who owned Marquar Cattle Station west of the Queensland border. John Morley was a city-bred man from south who was the new owner of Jervois Station. On a rather hot day they were both camped at Unka Bore attending to mustering the cattle. While they were having a few rums a friendly argument started about the cattle.

Suddenly Old John grabbed Mick's hat from alongside him on the ground and tossed it into the 20,000-gallon stock tank which was situated on Morley's property. That didn't seem to disturb Mick Brown very much at all. He calmly said, "I want that hat."

Morley said, "Well, you'll have to swim for it!"

Mick retorted, "No, I won't have to swim for it!" and slowly picked up his .303 rifle, put a bullet in the breach, aimed it at the bottom of the tank saying, "You bastard! I'm not going to get wet getting my hat out. If you want to save the water in your tank *you* had better hop in and get it out!"

Goat Duffing

When I closed down the mining and ore carting the second time, I left another caretaker out there to look after the goats, allowing him to use our living quarters. The chap was a drover, with a family of four, the eldest daughter being about fourteen years old, acting as 'mother' to the other children after their mother had died. He had been there with his plant of horses, in between droving jobs, but he wanted to go back to Queensland.

While we were back in the Alice he decided to move on after using up all the stores we'd provided. When I went out to check on things I found he'd been gone since about three months previously. The pump had broken down so he'd taken the goats and moved out on the stock route, selling the goats to Micky Brown near the Queensland border before he disappeared somewhere in Queensland. So that was the end of my herd of goats.

Some people may be critical of my always being willing to trust people at face value and I undoubtedly lost quite a large sum of money to cheats and liars over the years. There are always people on the look-out for ways to 'diddle' others out of their hard-earned money, but percentage-wise I can still say that I found more people who were trustworthy than the other kind. It wasn't in my nature to be distrustful and standing over people all the time to check up on them. An example of bush integrity occurred when we returned to Jervois after about a fifteen month absence to find a note on the table which read, "Helped myself to a 50 of flour, a bit of tea and sugar and some tinned meat. Will pay you back later when I see you." The storeroom was always left unlocked for occasions such as that if someone needed supplies and such people only took what they really needed.

After we went back to town my second marriage ended because of a great misunderstanding, with heartache for both of us. I carried on with my transporting business and mechanical work and married Elsie Dixon Collins later in 1958. Elsie became an instant mother to three adolescent boys! My only daughter, Paula, was born on 4th November, 1959 and another son, Kurt Damien, was born five years later on 28th November, 1964.

Young Kurt, Paula and Elsie at our Mount Charles home in the Alice, 1982

Chapter 39
Trials and Tribulations at Jervois

My introduction to the corporate jungle of company promoters, purchasing officers, 'legal eagles,' sharks and the straight-out smooth-talking rip-off merchants of the big cities left me utterly flabbergasted and very disappointed in the human race. I had been reared in the Outback where a handshake was as good as a ten-page legal document and probably much more reliable than one drawn up by an expensive lawyer with lots of legal jargon and terminology which needs a Philadelphia lawyer to understand it and can be twisted around to mean three different things.

This hit me like a bolt of lightning around 1960 after I approached the Commonwealth Development Bank for a loan to build a bigger, more economical copper sulphate plant at Jervois Mine to replace my pilot plant. The bank arranged for some engineers to go out to inspect the pilot plant, assess the situation and do a feasibility study. These two men went through everything quite thoroughly and were very impressed with my pilot plant which was producing about a ton a week. Their report was quite favourable and the Bank advised me that an advance would be recommended. It was suggested that the project might be a bit big and worrisome for me to handle and I should consider floating it into a public company, taking up a good issue of vendors shares and retire to live in comfort, leaving the every-day problems of management to someone else. They said it was one of the best projects they'd seen around and they were prepared to advance a loan of £30,000 to the company provided I would stand as guarantor.

I was told I needed a promoter to underwrite such a company if it could be proved it was viable. I was subsequently given the names of several potential brokers to contact. One of these was quite unrealistic as they wanted me to supply them with written contracts for forward orders for five years of 80% of our estimated production, and only allowing me 5% of the shares of the proposed company as vendors shares, with no cash settlement but a very large slice for themselves. Another wanted to over-capitalize the project by about 300%. The third one was acceptable to the Bank but it was suggested their proposal was under-capitalized, but further money could possibly be raised if required.

A few weeks later I was invited to Melbourne to get the float of the company organized. After floating the company I went out to Jervois and worked for about nine months building up a bigger treatment plant which I estimated would produce approximately one ton a day. However, things didn't work out quite as simply as I thought. Construction and the budget were on target and 75% completed when a dispute arose over me not supplying sufficient paper work for the head office in Melbourne. I told them I was too damned busy with the physical work of completing the plant rather than wasting my time writing unnecessary reports – if they wanted reports they could come up and do it themselves at their own expense! This didn't go down very well. The directors in Melbourne decided they'd send up a manager to organize the completion of the plant with written reports of progress at regular intervals. I was virtually dismissed as manager at that time.

They believed they would have it in working order quicker than I was doing. The new manager had some practical experience but mainly counted on his book learning from his university education and previous work under direction from a boss, to handle the job. In contrast I had never worked for a boss and all of my experience was based on practical application and solving of problems and I *knew* what worked out there. My frustration really built up seeing things deteriorate and not having any control over improving the situation.

By 1962 I was showing signs of strain and my doctor advised me to get away and take a break. I took a three-month holiday and went over to Western Australia to Rottnest Island with Elsie and Paula. When we returned the new manager had virtually run the company broke. Half a dozen extra men had been hired and the boss they'd appointed apparently never checked any of the time sheets. When I checked them I found some were very questionable – a couple of blokes had entered 23 or 24 hours of work and one even worked 26 hours in one day! Apparently the workers called that job the "Cream Wagon" – they just filled in their own time books and the manager sent away for their pay, all of which contributed to the collapse of the company.

Kurt, Paula and Elsie, 1964

When I'd left for my holiday there had been about £40,000 in the kitty, but when I returned the company was in debt to the tune of about £60,000. It virtually left me 'holding the baby' because I stood as guarantor to the Development Bank who had put up £30,000 towards the company. The chief director, who had employed the new manager, backed down and resigned as he was unhappy with lack of progress and the mess the whole operation was in.

Because of all these troubles the Bank foreclosed on the company. However, they said if I would carry on working my leases they would stand by me and not force me to sell up, as they had faith in me. Their faith was justified, as I eventually paid the loan off. Accounting for inflation, that £30,000 would be equivalent to about $200,000 today, which was a very large debt for me to repay but it saved me from bankruptcy. I accomplished this by mining the copper and shipping it directly to Mount Isa as the price had risen and they were allowing us to bring in larger quantities. I finished off a portion of the copper sulphate plant and had it working quite well again in a smaller way until the price dropped. I couldn't afford to finish the whole plant off.

A Medical Experience in the Bush

Shortly after my radio telephone was installed out at Jervois Range around 1965 I had a little episode of a medical nature. During the 1960's bomb shelters and radio telephones were installed by the authorities at Woomera for the safety of everybody living in the rocket path which extended over our area. We were then notified if rockets were to be fired in our direction.

I had been working back in the Alice at my Depot and had been hard at it all Saturday afternoon and Sunday morning, preparing my trucks ready to go on one of our last cattle transporting trips. I bought half a dozen bottles of beer for the boys to have a drink when we finished our work. While I was opening one of the bottles it burst and severed the sinew and one of the arteries in my little finger. I tied it up with a handkerchief and drove down to the hospital in town. The nurse asked about tetanus and I told her I'd never had a tetanus injection. She gave me an injection, stitched up the finger and off I went.

Next morning I drove out to Jervois on one of my trips in preparation for copper mining again. Tuesday morning when I awoke I felt strange and tingly and red blotches appeared all over my skin. By midday all the blotches had joined together, so I radioed that I suspected I had a reaction from the tetanus injection.

A plane was due out the following morning so the hospital personnel said they'd put some antihistamine in a package and send it out to me on the mail plane. At ten o'clock next morning when the plane landed, I was covered in massive hives running into each other like a big jelly and my throat had started closing over a bit. I collected my mail and asked, "Where's the parcel from the hospital?"

No parcel – nothing! It turned out that it had been put in the wrong bag! I radioed again and another parcel was sent out by car, but the car broke down on the way. It hadn't arrived by midnight and I was thinking of trying to drive into town myself. I even picked out a place in my throat to open with a sharp knife if my throat completely closed and I couldn't breathe. However, by about one o'clock in the morning it started to ease off and the swelling subsided so I remained at the camp. Later that morning I received three parcels of antihistamine, all around the same time, which I no longer required!

A Prospecting Foray into the Simpson Desert

Around 1965 I went on a prospecting trip into the Simpson Desert with Walter Smith and his wife Mabel. In earlier days Walter had accompanied explorers on camels through the deserts in Central Australia and Western Australia. He was one of a very well-respected half-caste family in 'The Centre' and Mabel was a full-blood aboriginal woman. As a lad Walter had lived for some time out in the Simpson Desert area with the aborigines. He told me he recalled a tribe out there who shifted camp from one rockhole to another where they dug up green rock, pulverized it and mixed it in water to kill the dingoes for meat. I thought the green rock might be copper carbonate and it was worth going out to try and find the place.

At the time we went out there Walter was around 75 to 80 years of age and becoming blind. Mabel acted as his eyes, with Walter describing what he was looking for. She would then look around at the hills and describe the terrain to him. The aborigines had a reputation for being very good at recognizing landmarks in bush country. We spent three or four days in that area in among the sandhills, about 200 miles (320 km) south-east of Alice Springs, but we didn't find anything. We found traces of rocks in the valleys between the sandhills, but it was obvious that during the 50 or 60 years which had passed since Walter had lived out there, the sandhills had moved quite a bit. He might have been off course too, not being able to see very well.

Our trip was cut short when we came across a herd of wild camels, consisting of one bull and half a dozen cows and calves. A wild bull camel in season can be quite dangerous, though perhaps not as dangerous as Mabel feared. She was absolutely petrified when the bull ushered his harem away and then advanced back towards us as if to warn us, "Don't you dare come any closer!"

Mabel spent a sleepless night worrying about the bull camel before we headed north and returned via the eastern side of the Harts Ranges and MacDonnell Ranges where we did some more prospecting. There were a few traces of molybdenite but nothing worthwile. I'd been told there was a place about 30 or 40 miles (48 to 64 km) east of Harts Range where the aborigines used the molybdenite to rub on their skins shining themselves up silvery black for corroborees.

Since we were running short of both petrol and water we made our way back through Harts Range to Jervois and came in around Mount Mary on part of the Indiana track. I dropped off Walter and Mabel at Valley Bore where they had their camp, before taking a look at a couple of other prospects in that area. I found a few small veins of scheelite and molybdenite and I also found some gem stones and beryl, but nothing large enough to be worth opening up. I returned again to Jervois where I was based.

'Bertha' crossing the Georgina River on the return trip from Mount Isa loaded with empty copper ore drums

Chapter 40
Joe Baldissera and the Turquoise

Joe and his Mail-order Bride
Joe came to Australia from Italy before World War II and settled near Tennant Creek doing prospecting and mining. Right after the War he left Tennant Creek and settled in Alice Springs doing general contracting work. Joe was ready to marry and settle down but hadn't been successful in finding an Australian girl to marry so he sent away for a mail-order bride from Italy. There had been very heavy rains and, on the day she arrived in the Alice, the Todd River was flowing fairly strongly.

Immediately after her arrival they were married and Joe took his bride to his home on the east side of the river. He romantically decided to make a quick trip back into town to get some flowers and wine from the shops, leaving his new bride waiting for him to return. Imagine his horror when he came back to the River Todd on the Alice Springs side and discovered it was in full flood! Poor Joe had to sleep in town and his wife was left on her own on their wedding night.

The Turquoise
One day early in 1967 I was in my Transport Depot Workshop when Joe turned up wanting me to fix his air compressor. I asked him if he intended going back into mining again.

"Oh.....," he said, "I don't quite know what to do. I've found something but I can't trust anybody. I think I might have to trust you."

Generally Joe could talk the legs off an iron pot, but now he was pretty close-lipped. All he would tell me at first was that we needed to get the air compressor and some other equipment fixed because he had found something that was "very good." He said he would take me out later and show me. The only ones who knew about it were Joe, his son and a half-caste man who had first shown it to Joe.

A couple of weeks later he returned, still hesitating and saying, "I don't know what to do!" He hummed and hahed and eventually decided that he would take me out about 300 miles (480 km) north-east of the Alice, near Ammaroo, and show me his find since he didn't know what to do about it or how to work it. When we reached this secret destination, his find turned out to be turquoise. The quality wasn't very high but there was plenty of it scattered around on the ground.

Joe didn't know how to attack the mining end of it and after some discussion we agreed that he would cut me in at a third, with one third each for him and his son. We would work it jointly with me supplying all the equipment which included a bulldozer, mining equipment, transport, camping gear and paying expenses. I agreed to that and returned to town.

I then went over to the Jervois mine and organised some of the gear I didn't need out there since the mine was closed down at the time. I also employed a couple of my best aboriginal workers and their wives who had previously been working for me on the copper. They were very good at sorting out the different grades of copper and turquoise slightly resembles copper.

Turquoise is a freak mineral made up of kaolin, phosphate and copper with each of these minerals occurring in the correct proportions to form turquoise.

While I was out there that first time I had a look around and found a good quantity of bone phosphate, which is a part of turquoise. It contains fish bones and fossils from the ancient shores of the Georgina Basin. I arranged for a consultant geologist, who was an expert on phosphates, to come up from Melbourne to inspect it.

When he arrived it was mid-summer and very hot. I took him there in the Holden utility, but little did I realise that he had a hole in his heart. He almost died on me out there! It was early morning when we first started walking around but in the heat of noon he couldn't stand the heat and collapsed. I built a little bough shelter and kept wetting his clothes to cool him down. I also gave him fruit juice and water to drink and eventually persuaded him to eat a bit to get some energy. After 3 or 4 hours he was reasonably recovered.

In the cool of the evening we had a look around before heading back to Alice Springs. I took him home via Jervois as he wanted to have a look at the mine there. We cut across to Ammaroo, then drove up the Sandover River and across to Jervois. It was a rough trip on a bush track, travelling most of the night to avoid driving in the heat of the day. Next morning we reached Jervois, had a look around, rested during the heat of the day and in the evening headed back to the Alice.

As it turned out, the high-grade phosphate rock was only in narrow bands. There was sufficient high grade ore out there but transportation costs were far too great for it to be a viable proposition. To market phosphate rock it has to be near the coast, handy to shipping or rail.

Anyway, back at the beginning of this turquoise venture, after gathering up all my gear and my helpers I went out there, put up a camp and started working. A few days later Joe arrived and I have never seen a man become so excited and raving mad! He said I had ruined everything by bringing out the extra aboriginal workers – everybody would know about it now! He raved on and on until about two o'clock in the morning. I couldn't get any sleep while he jumped up and down and thrashed around. Eventually I said in exasperation, "Look, Joe, if you keep on making all this racket the boys will hear you and will realise that it isn't copper."

He snapped out, "What do you mean it's not copper! Of course it's not copper!"

"Well," I said, "all the boys know is that they have come up here to help us mine copper. That is all there is to it. They don't know the difference."

I thought he had fainted or something because there was dead silence for several minutes. Then he came over to my bed in the dark, shook me first and then shook my hand, saying, "You have saved everything! That is a wonderful idea. I never thought of it that way." So that solved that problem!

Joe wanted every grain to be swept up all around the area where there was even just a tiny bit of colour. Some of it was more green than turquoise. He asked the women to pick up every scrap they could see under the trees and everywhere around so that if anybody came into the area or followed our tracks they wouldn't see anything. Once again I had to convince him that if he made such a fuss about it the workers would suspect it was something special. We worked for a couple of months and had quite a lot collected into drums. While I was away in town Joe carted it all over to the other side of the hill into a gully and smothered up the tracks. By this time the aborigines were awake to the fact that it couldn't be copper.

I was still back in town when the Asian flu hit which I caught and was quite sick for several weeks. All the while Joe was on my back to bring the road train out and pick up the 25 to 30 tons of turquoise we had accumulated by this time. It was mostly rubbish but he wouldn't leave one speck lying around. He thought every little grain, even smaller than a pea, was worth a lot of money. He had bought a brooch with tiny little turquoise beads in it, just a bit bigger than a pin head. He thought each of those little beads was worth about a dollar and was quite fanatical about saving every little piece, even if it was only a pale ashy-blue colour.

When I drove out with the road train I was still unwell and taking antibiotics. Joe wouldn't accept any other driver going out – he didn't want anyone else but me because others might wake up to the turquoise find. We loaded it up, but I had a relapse on the way back about 60 miles out of town. I felt so crook I simply had to make camp and take a rest at about 2 o'clock in the morning and completed the journey to the Alice next day.

Joe became all excited again, raving away at me because he'd wanted to arrive in town and unload the truck overnight so that nobody would see him putting the drums of turquoise in his shed. Again I said, "Look, Joe! For Chrissake! If you make it secretive like that, everyone will wonder what is going on in the middle of the night! Best to do it in broad daylight and just say you've got some copper ore coming in."

Said Joe simmering down, "Oh yes, that might be a good idea." So we did that.

Several days later, while working alone in my workshop on a Sunday afternoon, I felt dizzy and collapsed. I lay down on the floor for a while and then drove home. Elsie put me to bed and called the doctor. It was my first heart attack which put me into bed for four or five weeks. I went south to see a specialist who confirmed the heart attack and during my recuperation I was advised to go on a trip and get away for a while. This was when Joe asked me to market the turquoise. He gave me a handful of specimens (about half a pound, I suppose) and I was to travel all over Australia, to Noumea, Hong Kong and Japan and try to find markets for the turquoise. (Some recuperation trip!)

I decided I couldn't very well go marketing the turquoise with such a small, inferior handful of specimens so I 'pinched' about 30 kilograms of different grades and took that with me. I left little samples, about a quarter to half a pound, with some of the bulk dealers.

A few weeks later Joe went south and spoke to one of these dealers, showing him a tiny piece about as big as a plum seed without knowing that I had already negotiated with this chap. Joe was making a real fuss about this tiny piece and wouldn't let the dealer have it. When the dealer opened his drawer and pulled out a piece as big as my fist, Joe nearly had a heart attack too! Several months later, when I saw the dealer again he was still chuckling about Joe.

Anyway, that episode made Joe terribly suspicious of me – he thought I must be giving fortunes away while I was over in Japan. He went to Sydney and contacted a mining firm there and eventually had an offer from them on paper of $2 million for the whole lot, subject to so many tons being mined out and so forth. It all boiled down to him eventually getting about $200,000 to $300,000 for it, which was a very good price.

When I returned from my holiday he came to me wanting to buy me out without telling me about the offer he'd received. I had my suspicions that he must have foxed out a buyer somewhere, but I was fed up with him and said, "Yes, you can buy my share any time you like."

He asked, "How much do you want for it?"

I said, "Well, I've spent around $18,000 on wages, explosives, bulldozer and general expenses. Double that will do for my own expenses, so I suggest $36,000." He harped on a bit about it and wanted to pay me by instalments so that he wouldn't get taxed too heavily. Eventually I said to him, "Look, I will settle for $25,000 – $13,000 this month (June, 1968) and $12,000 next month (July). He finally agreed on that and I was very glad to get out of that show.

Quite a good quantity was subsequently mined out from there. The company was called International Turquoise. John Cummings worked it off and on for quite a few years. However, that was the end of the turquoise mining venture for me!

A few years ago in the 1980's, poor old Joe stepped out from behind a car in one of the side streets of Alice Springs and was killed. He would turn over in his grave now if he went out there and saw how the turquoise is mined these days. Low-grade turquoise is scattered all over the place in great big heaps visible to anyone who might pass by, with only the better quality rock being taken.

Chapter 41
Return to Jervois

South for a Spell
Using the money I received from selling out my shares in the turquoise mining, and adding the amount received from selling my shares in Jervois Copper Sulphate on the open market, I went south for a spell. There had been an attempt to block me from selling those shares on the grounds that I had been a director of the company, but the Stock Exchange allowed the sale since I had long since resigned by that time.

I bought a house at Holmes Court, Flagstaff Hill, South Australia, on speck and started building another one at Fama Court, Flagstaff Hill. Elsie went to England to visit her relatives while Paula and young Kurt stayed with my brother Randle and his wife Ruth. When we moved into the new house Randle and Ruth lived in the Holmes Court house.

In 1970 I returned to Alice Springs to get my transport business going again, as the manager whom I had appointed in my absence had upset the stock agents and some of the station owners. Elsie didn't like the idea of me working in the Alice while the family remained down south so we let the Fama Court house, which was later sold, and all went North again, back to the Alice. However, my interest in the cattle transport virtually ended at that time.

Back to Jervois Mine
We lived in Giles Street, Alice Springs while I built a house on my property near Mount Charles, north-east of town. We moved to Mount Charles early in 1971 while I was preparing to go back out to Jervois Mine, where I still owned my original leases. The drought had broken at Jervois and copper prices were up so I'd decided to try my luck once again out there. I had some new equipment which included a Crawlair drilling plant and 922 Caterpillar front-end loader. My oldest son, Lindsay, was selling the ore to Mount Isa.

Around 1970 I was referred to Bryan Fitzpatrick in Melbourne, who became a true friend. At the time I was really frustrated as I thought I'd be forced to sell up and get out from Jervois at considerable financial loss. I decided to phone Bryan, who owned Petrocarb Exploration, and was given an appointment for half an hour the following Monday, so I went down to have a chat with him. He'd already heard one version about what was happening at Jervois. My half hour with Bryan lasted a full two days and it turned out he had his own very good reasons for being willing to listen to me. Bryan subsequently agreed to buy my leases which were later transferred over to Petrocarb Exploration, which later went broke owing to a collapse in metal prices and other difficulties.

Finishing at Jervois Mines
Several years later, around 1979, Bryan was overseas working on the London Metals Exchange and obtained some foreign capital from a company willing to invest in the Jervois mining leases.

Around $14 million was spent putting up a big plant and developing the mine. The new company was called Plenty River Mining (P.R.M.). It was intended to call the mining township Johannsen to honour my work there, as mentioned in the August 1980 issue of *New Outlook*, House Journal of the Colonial Mutual Life, but the township never eventuated.

Just as they were about ready to start the mill running, after test runs were actually done and there had been an opening, world copper and silver-lead prices dropped so badly P.R.M. had to close down, otherwise the mine would have been worn out without making any profit. Some small development with a limited crew of workers was continued for a while, waiting until prices might rise, but eventually it was left under care and maintenance.

Recently, in 1991, it was sold under auction and all the buildings and equipment dismantled. The area has now been taken over by Poseiden.

The truck returning from Mt. Isa with empty drums. Note the self-tracking trailers.

Chapter 42
Bonya and Molyhil Mining

Around 1972, Lindsay did some prospecting at Bonya Hills 20 miles (32 km) east of Jervois with an aboriginal lad called Alfie Toby. Actually, Lindsay had been settled at Bonya for quite some time prospecting around the area and he'd found some fairly rich scheelite deposits not far from his camp, where he had a caravan and a couple of sheds.

Central Pacific Minerals also had quite an extensive prospecting camp near there, employing about twenty men. They had become very interested in Lindsay's finds and wanted to make a deal with him to purchase them. However, every proposition that was put to Lindsay had too many strings attached. Central Pacific Minerals' head office in Sydney always seemed to have some reason for not agreeing to Lindsay's terms or the purchase agreements were reworded making them completely untenable as far as Lindsay was concerned. The bottom line seemed to indicate that Central Pacific would gain practically everything and Lindsay would get very little.

This kind of hassle continued for about six months with Lindsay slowly going broke but he was still hoping to either work it or sell it, as the prospects were quite valuable. Scheelite prices by then were quite high, since it was in great demand again. Eventually Lindsay approached me and said, "Look, I've found all this scheelite but it's too big for me to handle alone. I can't work it and whenever I try to sell, it seems as if Central Pacific Minerals are trying to beat me down. What about you coming in for $10,000 and we'll go 50-50 if you can market it or work it with me jointly?"

I agreed to do it. At that time I was selling Jervois Mine to Petrocarb and had about $40,000 or $50,000 coming from them for plant, equipment and so forth. In exchange for a half share with Lindsay the first thing we did was to build a house for him and his wife Joan, using the same plan as my own house which I'd finished building just out of Alice Springs. Lindsay and Joan named their home "Baikal".

I hired some contractors to put down a couple of bores so we had one near one of the mines and another at Lindsay's place. I then negotiated with Petrocarb to take an option to buy several of Lindsay's scheelite prospects at Bonya, giving them the right to exercise an option to take up any portion of the option lease by lease. The main lease was priced at $100,000 which they exercised, enabling Lindsay to pay off his debts. We formed a family syndicate to establish ourselves in mining Bonya and Molyhil.

Using several aborigines to help us, we did some further prospecting with Lindsay doing the main part. We supplied Alfie Toby with a Toyota land cruiser in payment for the prospects he found for us and staked him with supplies and petrol for any further prospects. We also paid him a wage besides giving him bonuses for anything worthwhile that he found. After showing him samples of the rock types with which the scheelite was associated. During the day he'd attend to the geological aspects of searching for suitable sites and at night he went out prospecting using an ultra-violet lamp which revealed any scheelite amongst the rocks.

The aborigines were very good at prospecting. We employed them quite frequently at Jervois and Molyhil. Lindsay and I found three good places while the aborigines found about thirty different spots which contained scheelite. Whenever they found a speck they would show it to us and ask, "Might be good one?" Then we would go and have a look. By this method twenty or more prospects were found, most of which were only small and too low a grade to be viable. Luckily we were searching during a drought period when there was little vegetation on the ground, which was suitable for walking around in the dark, though we sometimes received a poke in the face from a dry stick.

After about six months we found a good prospect at Molyhil which was about 20 miles (32 km) further west from Bonya on a very bull-dusty, sandy track through the scrub and several heavy sandy creek crossings. Alfie had found traces around there and the Bureau of Mineral Resources had also found traces of molybdenum on top of the hill but they'd never found any scheelite or extensions to it.

When we first prospected in the area and pegged it, I went into town to register the leases and discovered that a company from New South Wales had taken out an exploration licence over the whole area which meant that we couldn't register any leases on it. We had to just "sit quiet and say nothing", taking our pegs out again. This was during the big mining boom and people from all over Australia had been applying for exploration licences in areas where many of them never even went near. When the big collapse came these areas were simply abandoned.

In the meantime one of our aboriginal boys, who had previously been employed by me, went over to work for a chap called Ramsay. Ramsay and Bluey Bruce were working on a copper show about 20 miles (32 km) north of us at one of my former prospects. They'd jumped the lease because it wasn't being worked. I'd already sold that lease to Petrocarb and since they had an exemption on it there was a court case. Bluey and Ramsay had pegged and registered it, but they'd described the location incorrectly. Anyway, their lease was 'squashed', but it was another one of those niggling things which wasted our time and money.

Just after the court case, this aboriginal worker told Ramsay he knew "where Johannsen got some good prospects for scheelite." The New South Wales Company still held the exploration licence for the area at that time, but for a couple of flagons of wine this bloke took Ramsay over and showed him. Ramsay immediately pegged the area and applied for it. Of course, his application was knocked back by the Mines Department because the New South Wales company still held the ground. We were very lucky it was thrown out of Court at that time.

Later on I was in the Mines Department to ask them about these N.S.W. people and whether they intended doing any work in that area. Under terms of an exploration licence, exploration work must be done within certain time limits as part of the covenant. Anyway, the Mines Department said the Company hadn't been doing any work out there and they'd just received a letter from them a couple of days earlier cancelling the exploration licence. They showed me a copy of the letter from the Mines Department notifying the company that the cancellation had been granted. That was at about 10 o'clock in the morning.

That afternoon, if anyone was watching, a black streak might have been seen all the way up the North road where I was tearing out to peg these areas again! I was within about 4 miles (6 km) from the area when I came across some fresh motor vehicle tracks. There had been a few showers of rain earlier and here was this motor vehicle track, just after the rain, only a couple of hours old! I thought to myself, "Oh hell! Ramsay's probably found out about it and gone up and pegged it again." About half a mile off the area I was relieved to find that the tracks veered off. It was probably a station vehicle which had been looking around to see how much rain had fallen in that area. I immediately got busy with the pegging. A few more heavy showers came down but I didn't stop. Since I had measured the boundaries before it was quite quick work. All I had to do was put down new pegs, put the papers up, go back into town and register them.

About two weeks later Ramsay found out that the area had been cleared. His idea had been to try and get the leases granted before it was cleared so that he'd get first preference. However, it didn't work that way and he landed in the Warden's Court at the hearing of the leases. He made objections, but we had witnesses there to prove that we had previously pegged it and removed our pegs, so he lost out and wasn't very happy about it. We also brought in the aboriginal who'd gone over to Ramsay. He wasn't very happy with Ramsay either, because Ramsay had promised him two flagons of wine and two more later on but had reneged on it. He never got his wine so he was a witness against Ramsay.

I then went back out with half a dozen of our working boys from 'Baikal' and a couple more from in town. We cleared the boundaries properly and started digging. Unfortunately, I got a bad bout of the 'Hong Kong flu' and was really sick, out there in the middle of summer, sweating like blazes. At the time I thought I might end up being buried out there. I had a few aspirin tablets with me and I asked the boys to make a shade for me and bring me up a bucket of water and a pannikin. I had a sheet in my swag which I put over myself and kept wetting to bring my fever down. The aborigines were certain I was going to die and started clearing off to the other side of the creek about half a mile away as they didn't want to be near me. The aboriginal tradition if someone is dying, is to leave them and shift camp. They reckoned the 'kadaitja' might have pointed a bone at me since it was near one of their Dreamtime rain-making places and their sacred caves were only about 4 miles (6 km) away. One of them continued to bring me more water and after about three days my fever abated and I was well enough to drive back into town. I then recovered fairly quickly and a few weeks later Lindsay and I moved the caravan out there and started mining with our aboriginal workers. Our families remained back at the home bases.

Lindsay and I thought that the scheelite prospects could possibly develop into something worthwhile. We couldn't find out where the 'shed' originated but there was an area of about 20 acres on the surface which had crumbs of scheelite scattered all over it. We searched around several times and one day I said, "Well, look Lindsay. Let's get a couple of the boys and start them off digging down in the most likely spot to see if we can find anything further down where this came from." I took a pick and drove it into the ground and said, "You can start digging a hole there."

When I lifted the lump of dirt out where I'd driven the pick in, it turned up one of the richest patches in that area! Boy oh boy! From then on we found quite a lot in big blobs around Molyhil itself. At first we were picking up bits about as big as plum seeds. We cleared the scrub and had aboriginal women and children foxing around picking up bits and pieces to see if they could spot any richer 'sheds'.

Lindsay, Dominic and Kurt Johannsen (father, son and grandfather) and Damien Miller (second from right) at the Molyhil camp. (Photo courtesy Bryan Fitzpatrick).

We three Johannsen men (father, son and grand-son) made camp under a couple of gidgee trees, just pushing some boughs in the top for a bit of extra shade. It was midsummer and stinking hot with temperatures around 45 degrees Celsius. We had a kerosene fridge, a couple of army stretchers and our table was a door stuck up in the fork of a tree with a prop holding it up on the other end. That was our camp, which shows how we started off in quite primitive conditions!

Gradually we developed it and looked for a market. Tennant Trading in Sydney were interested but advised us that if we wanted to get onto the international market for a good price we'd have to supply a minimum of 15 tonnes of high grade scheelite, which is a minimum of 65% tungsten, to become a recognized world-producer and qualify to receive the world market price. Otherwise we would only receive the lower price given to small producers, about two-thirds of the normal price. To qualify we had to supply the first 15 tonnes at a 50% discount. At the time scheelite was worth around about $4,000 a tonne whereas a few years later it increased up to around $8,000 a tonne.

We busied ourselves with the machinery and employed about eight aborigines working on hand-sorting and scratching around in little pits in the richest places, after which we'd jig and clean it. Eventually we accumulated our first 15 tonnes and sent it away. In the meantime we'd already drawn a deposit on that tonnage from Tennant Trading to keep us going and used all the funds from that sale for equipment, wages and other expenses.

By this time we'd put the caravan in a big bough shed, which made quite a pleasant campsite alongside the creek. We didn't see any signs of flood marks around, although we suspected that on high floods there might be 6 to 12 inches (15 to 30 cm) of water going through there. Occasionally Elsie and the children came out to stay for a while during the school holidays and I went back into town every fortnight for stores and fuel. Once we had moved into the bough shed and the caravan things were a bit more comfortable and I brought some of our mining equipment over from Jervois and Alice Springs. I'd bought the Crawlair percussion drill back from Petrocarb when they closed down for a while. I'd sold the drill and compressor to them earlier for $20,000, but they wanted the same price back when they closed down, which contributed to draining our bank balance.

We put down over twenty percussion bore holes looking for water but it was all so putrid we couldn't use it. It was like acid, full of fluoride, salt and other minerals, which wasn't even fit for use in the plant. We had to cart water from Marshall Bore, about 10 or 12 miles (16-20 km) away. I borrowed an old practically worn-out bulldozer tractor from Petrocarb at Jervois. With a great deal of struggling and battling we put down a catchment dam about a mile away from our camp and gradually improved the treatment plant. We accomplished many things by experience, trial and error and lots of hard work!

In the meantime, after we'd sent our first consignment of scheelite away, the price dropped to about half and instead of receiving a total of about $60,000 for it, less what we'd borrowed, by the time the 50-50 basis was cut out and all the other factors, we finished up receiving only about $8,000 with which to carry on operations. We battled on and for the next consignment we received the full price. By then the price had increased slightly to about $2,500 a tonne which just about covered our expenses. Everything was tooled up to plough the money back in so we could carry out further development and put in a catchment dam for water. Rising and falling prices create a miner's nightmare in terms of planning finances.

I worked in town for a while building three concentrating tables. Since we didn't have a power plant out at Molyhil I motorised these tables using 12-volt generators from cars and windscreen wiper motors. I used my welder for a power plant with little belts running in all directions. We also rigged the crusher up and were producing quite good concentrate, but it was a slow process. We were only producing about one ton a week even when things were good. We'd work steadily for five or six weeks then we'd have to modify or repair the plant as well as keeping up the mining.

By this time we were receiving a better price for consignments of ore but we were always between $50,000 to $100,000 behind, having to get advances from the buyers and the bank. Luckily 'Knocker' White in the National Bank bent over backwards to be helpful, which made our situation much easier. At that point I was having problems convincing the family that it was all worthwhile when every bit of production profit was being ploughed back into the mine for equipment, housing and so forth. Because it was such a worry, they wanted to get out of it – to sell the next consignment and sell the leases to the first bidder. Lindsay and I had been giving them long chats or a 'pep-talk' every month, trying to keep their spirits up, encouraging them to carry on. We'd explain that if we were buying a house on a 20-year term they'd be quite happy to keep on paying interest and paying off a bit on the principal each month. We'd try and convince them that it was a good proposition and that we were developing it for a family investment which should eventually work out quite well, but they wanted to get out as they'd had enough.

Primitive but very effective treatment plant at Molyhil (Photos courtesy Graeme Mudge)

However, we continued on for the time being and received some assistance from the Government. They put in a few Diamond-drill holes for us and proved up more ore – we finished up proving about 1,500,000 tons of ore. Prior to this we had uncovered some more extensions by stripping about 3 feet (1 metre) off the surface soil. Underneath was a great long band of beautiful ore which really boosted our production. That ore band on the surface was drilled with a Diamond drill down to about 600 or 700 feet and it was still showing ore, which proved it was a good prospect in scheelite and molybdenite.

We never recovered any of the molybdenite because we couldn't afford to put in a suitable treatment plant to recover it. We had hoped to accomplish that once we got our heads above water. The molybdenite in the ore was worth just as much, or more than the scheelite we were already extracting, so we believed we were on quite a bonanza. However, our financial problems continued and the women still wanted us to get out even though they were beginning to believe it could possibly be a really good thing in the future.

We kept on working at it for the time being and were going along pretty well with a workshop set up, a roof over the treatment plant, a couple of extra concentrating tables, several power plants and a bigger crusher. We had also started to open-cut the mine and were just about breaking even. When I say "breaking even" – each consignment was worth about $40,000 or $50,000. By the time the next consignment went out the wages and costs would be about half that amount. We were reaching the point where, if we had a clear run without major breakdowns, in about four or five months we expected to be clear of debt and each consignment would be showing a profit of about $30,000 or $40,000, which looked really good.

However, the worries of finance and so forth were taking their toll on my health. I was over at the dam fixing up the pump one day around 1977 when I suddenly felt quite queer. To cut a long story short, it was a second mild heart attack. I was flown out by plane and put in hospital for a couple of weeks. Lindsay tried to carry on but it was too much for him to take on the responsibility of mining and all the other operations. His wife Joan was also sick at the time and he was travelling 32 miles (50 km) each way every day so she wouldn't be alone at night. She couldn't stay at the mine because their home had to be looked after.

Selling Molyhil

The family syndicate decided it was time to sell out, so we contacted Petrocarb and organised the sale of Molyhil on a time payment basis. They took over immediately and started producing. At first they planned to modify my plant and update it. They reckoned, "If Johannsen could produce a tonne a week they could more than double that by modifying the plant." After spending about $6 million on plant, wages and generally wasting of money, they never even got up to a tonne a week. When the price dropped they had to close down and eventually sold all the equipment and housing by auction around 1986.

Before Petrocarb took over, things had been really getting up to scratch at Molyhil and we'd installed power lines to two demountable homes which we'd brought over from Jervois. Setting up a mining camp like that costs a lot of money, but once we'd proved the value it was a case of getting the most out of it that we could. Before we handed over we had started on the new open-cut so there were thousands of tonnes stacked up ready for treatment. The area was further drilled and doubled the amount of proven available ore. When they first took over there I thought Petrocarb would simply upgrade my plant a little and use it. They could have used it as a pilot plant to find out to design a bigger treatment plant. However, they were confident they could upgrade it on a much grander scale with their engineers and experts.

When Petrocarb closed down quite a lot of the plant and equipment was sold off. Earlier, there had been a verbal agreement and some items, such as the drill and big front-end loader Euclid, had been promised to us. Jack May, one of the original directors, had agreed to give us first option on buying that equipment in return for our promise that we'd give them first option on any prospects we found. However, there was a change of directors and verbal agreements had been made without written verification. Jack May became ill and died and the Head Office refused to honour any promises Jack had made. Of course written agreements should always be insisted upon in any business dealings, but that wasn't always the way of folk in the bush.

One of my road trains ('Bertha') had been left for Petrocarb to borrow until they got their own equipment. It was used for a while and a few parts on it broke down. Next thing one of their air compressors also broke down. Since my road train had a similar motor in it they pulled half the motor out of that to keep their air compressor going and left the rest of my vehicle open in the rain and weather. When I had recuperated from my heart attack and returned, wanting my road train back, they just said, "Oh, we thought Petrocarb owned it, but we never gave anyone authority to use it or pull parts of it. If you want it, take it as is or we will give you $2,000 for it." I left it there for the time being, but 'Bertha' was later retrieved and now resides in Alice Springs Ghan Preservation Society Museum where it is being restored to eventually be housed in the proposed Road Transport Hall of Fame.

Petrocarb sold off other equipment without giving us a chance to make a bid on it. A lot of it went up to Darwin to a dealer for next to nothing. Yet the company still owed us about $1,400,000 at the rate of $1,000 a week split four ways to Lindsay and Joan, Elsie and myself. We finally decided to cut our losses and accept an offer for final settlement and split it four ways.

Close Shaves at Molyhil

We were very lucky while working there in relatively primitive conditions, that we never had any really serious accidents. However, one Sunday afternoon I heard a different sound coming from the mill when two of the aboriginal workers were running it. We had some visitors, including four young children ranging from five or six years up to about twelve. The children asked, "Can we come over with you?" I said, "Yes, as long as you stay in the back of the ute." I had a 1-tonne Holden ute and the smaller children stayed in the front seat while the bigger ones rode in the back.

I pulled up alongside the mill on the side of the hill where there was a ramp. I shut off the engine, put the ute into first gear and ran over to the mill to check it. I think the power switch was off or something like that. I fixed the problem and was just on my way back up the hill when one of the kids screamed out, "Kurt! The car's running away!"

I ran a helluva race up to the car. It was rolling down the hill and heading straight for a big tip truck standing at the bottom of the rise. I thought the children were still in the car, but actually, as they had got out, one of them must have accidentally knocked it out of first gear. They weren't in the car at all – they were alongside of it, but I didn't know that. The driver's side door was still open because I had only jumped out to do this little job. I raced down the hill after the ute and managed to fly into the open door with my left foot trying to slam on the brake. The first time I missed but I managed to get one hand on the steering wheel with my other leg dragging outside. My right knee skidded along the gravel which tore my kneecap to shreds. There were great gashes all over it, bared to the bone. I finally managed to get my foot on the brake and skidded to a stop within about 6 feet (2 metres) of the truck. That was a close shave and I was laid up with a 'gammy' knee for about two months before it healed. If the children had been in there and it had slammed into the truck all of them may have been badly injured.

Another accident which could have been serious occurred when one of the aboriginal fathers was working on the crusher. His young daughter, about eight years old, went over to the conveyor belt, which was an area supposedly completely out of bounds to all the children. Her father hadn't seen her standing there behind the belt tightener, which was a double set of pulleys to keep the conveyor belt tight. She was rubbing her fingers on the belt as it went under the pulley when suddenly her arm was taken in under the pulley. By some freak the belt stopped. It must have had an extra heavy load up on top and the drive belt at the motor had slipped. She ended up with a very sore hand out of it but no bones were broken – just the skin broken. It made me shudder to think of what may have happened if the conveyor belt had been empty – she may have had her arm ripped off.

Yet another accident occurred when one of the really old tip trucks was carting ore. It was backing up to the ramp to tip a load of ore into the big ore bin. The chap who was driving it possibly didn't put the truck into gear properly. The open cut was about 50 feet (16 metres) deep only 50 metres ahead of him around a gradual curve on a big ramp. The truck suddenly jumped out of gear, but instead of steering it down he panicked and jumped off, letting it go. Luckily he landed safely, because the truck hadn't yet picked up speed. The truck landed on its nose at the bottom of the open cut mine, which was the end of that truck! It was a complete wreck. All we saved from it was the motor, which somehow escaped much damage, and a few of the tyres.

With machinery you have to be on your toes keeping parts oiled and greased and seeing that the workers do things correctly. Wherever there is machinery in constant use, something always needs attention.

Part of the open-cut mine at Molyhil (left) and the crushing plant (right).
(Photos courtesy Graeme Mudge)

Chapter 43
Droughts, Floods and Fires

Living out near the edge of the Simpson Desert, there were always special challenges, dangers and frustrations. However, there was also a wonderful sense of exhilaration and freedom that those who live and work close to the beauties, bounties and blows of Nature know. Droughts, floods and fires were all tackled in those years before we pulled out of Molyhil and Bonya.

An ever-present problem at Molyhil was the terrible shortage of water for treatment of the ore. I first decided to use a rather small pit into which we pumped the slurry to settle out. We fenced it off because Johnny Turner, owner of nearby Jinka Station, had a prize bull which was inevitably attracted to the lovely green grass around the settling tank or pit. One day, before we fenced it, the bull decided to walk into the slime pit for a drink. It was quite a valuable beast, but if he had drunk the water in the settling tanks it would have been the end of him. We hunted him away to one of the other bores and fenced it off as soon as possible after that incident.

When it became necessary to increase the size of our settling tank we built up a sizeable dam about 150 or 200 feet across and carted about 300 tonnes of good-quality clay from the catchment dam which had been put down earlier. We compacted about 6 inches (15 cm) of this clay all over the bottom of the new dam and up the walls to seal it. When we tried it out it worked perfectly and held water like a bottle. I put another bore down, which must have been about the seventeenth bore we'd drilled in our search for a decent water supply. This one gave us about 200 gallons (910 litres) an hour. By pumping 24 hours a day we obtained a good quantity of water in the new settling dam, but the quality was putrid. Also, we didn't realize that the acidic fluoride in the water was breaking the clay down and within a week the clay seal no longer existed. It had become mud and the water disappeared through it like a sieve. All that hard work for nothing!

During that same period several scientists were operating in Alice Springs doing stratosphere balloon testing. A balloon sent up into the stratosphere was lost due to a leakage and came down about 30 miles south-east of our place. Lindsay noted the direction of it, went out looking and eventually located it. The balloon party were also searching for it so when Lindsay notified them he had found it and showed them where it had come down, they collected the pay load which was worth many thousands of dollars, but said we could keep the damaged balloon material. It was made of about 6 acres (2.4 hectares) of very thin but durable clear plastic.

Lindsay and a couple of helpers cut it into sections, gathered it up in huge sheets about 20 or 30 metres wide, folded it and loaded it on his truck. It squashed down pretty well. We figured we'd find a use for some of it, and decided to lay this plastic material on the bottom of the settling tank, cover it over with tailing sands to protect it, thus making a seal in our buggered-up tailings dam. It was great! It worked beautifully until Johnny Turner's old bull decided to come back. He wasn't content to just walk in and get himself a drink. He walked up and down the bank, criss-crossing everywhere and punched holes all over and through it! So that was the end of our tailings dam – again! Such things are sent to try us! Water was so precious and prior to putting in the catchment dams the drought had caused the dam to remain dry for months after completion.

Big Rains!

Around 1976, a couple of years before we sold out, we finally got rain. It rained and it rained and it rained. About 200 metres from Molyhil, calcite and fluoride crystals can be found, which is one of the aboriginal legendary rain-making places. According to their traditions they smashed up some of the rocks into crystal fragments and threw great handfuls up into the air. Glistening in the sun, it resembled hail coming down, which was part of their rain-making ceremony.

Anyway, one day during the drought, a couple of the aboriginal men and I were there fencing the area off so that other people couldn't come in with a bulldozer or other machinery and damage the area. If that happened there'd be hell to pay with the tribal people. I jokingly said, "I think I'd better make some rain." One of the aboriginal boys said, "Oh, I've bin tryin' for weeks, but I can't make any. Only that old fella Blue Bob was rain-maker and him bin 'pinish' long time. You better have a go, boss!"

I gathered up a couple of handfuls of the crystals and threw them up into the air and executed a bit of a dance. Blow me down, a few days later, down came the thunderstorms! After we'd had about 10 inches (250 mm) of rain the aborigines designated me to be their new rain maker. That year we had 43 inches of rain, when the normal rainfall for that area was about 10 inches a year.

Near our caravan underneath its bough shed, Molyhil Creek began to flow. Another great thunderstorm rolled in so I said to my workers, "I think we'd better move the caravan and gear to higher ground. The creek's starting to overflow." Close at hand, about 20 metres away, the ground rose about a metre so we parked the caravan and most of our gear up there. Before long the creek flooded over where the caravan and bough shed had previously sat. Still it rained and rained – and rained some more. Next day as the rain continued the creek rose higher. There were more big thunderstorms with deluges of rain and the water continued to creep up, lapping around the wheels of the caravan even where it was located on the higher ground. I thought, "Well, I'd better move all this gear even further up the hill." I did this by taking one load up at a time and returning for another. By the time I'd get back to where I'd left the first load of gear, the water had risen even higher so I'd have to shift it all again!

Ahead of the quickly rising flood waters were thousands of frogs all heading up the hill. I thought to myself, "They must know something...." so I shifted the gear higher still and raced back for the last few bits. By then the water had risen to floor level at the steps of the caravan. Realising I needed to act quickly, I sprinted up the hill where I had a brand-new roll of about 100 feet of one-inch green plastic garden hose. I quickly cut the strings on it, unravelled it and tied one end around the caravan tow bar. There was one fairly big gidgee tree to which I managed to attach the other end, getting half a knot on it around a limb. That was as far as I could knot it so I used a piece of wire to prevent the knot from coming undone. Only minutes later the caravan rose on the flood waters, floated and swung around, then settled down again. The water continued rising until it was within about half an inch of the double-bed mattress inside the van.

Finally the water started abating, leaving the van full of mud. My next task, once it had sunk below the door and floor level, was to start bailing the water out as the flood receded. A couple of our aboriginal workers, Norman Reiff and his sister-in-law, Emily Ben, came over to see if I needed a hand. They had shifted camp into the workshop further up the rise. There we were, bailing water out of the open door and sloshing it around before the mud settled down too hard. We washed, scrubbed and cleaned out the caravan while the water was still running underneath it. The floods rose again several times, but not quite as high. After that the aborigines jokingly gave me "the sack" as rain-maker because I'd made too much and didn't know how to stop it! I was jokingly banned from going near the rain-making area!

We couldn't get back into town for a while because all the rivers between Molyhil and town were flooded. We were stranded there for about two weeks before we could get across the Marshall River. The Plenty River was still too high so we waded across, up to our necks and reached Huckitta Station where Quentin Webb lived. He agreed to kill a beast since they were also running short of meat. We carried our half, cut into smaller pieces, across the Plenty River on our heads which gave us enough meat plus our dry rations to last us a while.

There were four major rivers to cross to get into the Alice – the Thring, the Plenty, the Marshall and the Entire. Then it rained again! A couple of weeks later the rivers had subsided enough that we managed to cross the Plenty River with the Euclid (the big front-end loader with its big tyres) and sent a radio message to Harts Range from Huckitta to get some stores sent out to us. The Entire Creek was still flooded with the concrete crossing completely washed away and the river too deep to cross by vehicle. We waded and swam across with the stores and managed to get back across the Plenty.

On another trip when Lindsay was running short of stores, we went to the eastern crossing near Jervois station which was impassable to vehicles. We carried the stores across there and also across the Marshall. Lindsay tried to negotiate the river with the FWD with its big 4-feet high wheels but he only managed to go in about 50 feet when he came to a big wash-away and drowned the engine. He used a grader on the other side to pull the FWD out. Everything had been flooded with water and had to be drained. By this time, after months of heavy rains, the ground was really soft and soggy. We could hardly walk on the roads, let alone put a motor vehicle on them. Each time there was a shower of rain everything would start running again. I don't think there was a month that year when we didn't have some rain. Those days were exciting, hard-working times, but good really, because rain is always appreciated in that country, even if it does sometimes arrive in quantities that are rather too large!

Robby's Bogging Escapade

In between rains I left a young chap, Robby Warner, out there for about four weeks while I went back to the Alice. He was only about 18 years old at the time and a bit of a harum scarum, but had plenty of initiative and was a good hard-working lad. There were also two aboriginal boys out there with him. They had plenty of stores and everything they needed but they became a bit bored with the continual rain and little to do, so they decided to come into town, which of course was almost impossible. This misadventure had started when they opened a window into my room in one of the demountables and found a couple of cartons of beer and a carton of wine flagons under the bed. When I was out there I always rationed the grog out to the workers. Anyhow, the three of them got stuck into this grog and got themselves nice and "plastered". When they had finished it they made an inebriated decision that they'd have to go into town and get some more. Any excuse!

They set off with a 4WD Toyota and managed to travel about 6 or 7 miles (10 km) but it was hopeless trying to wallow through the mud. They became firmly stuck and couldn't shift it at all so they walked back to the camp and decided they'd take 'Bertha', the prime mover of the road train, to pull the Toyota out of the mud. They were about 50 metres away from the Toyota before 'Bertha' also became well and truly bogged. Undaunted, they walked back to camp again and took the big D7 bulldozer. With the optimism of youth and alcohol, they drove the bulldozer out and also bogged that! After walking back to camp again they gave it up for the day.

A day or two later they decided to try the Euclid, but in the meantime it had been raining again and the road was even more sloshy. However, they pushed ahead with their plan and found they could go along reasonably well in the creek bed where the water was running and the sand had washed right down with rock underneath. They travelled along in the creek bed for a while but couldn't get up the sides to get out – the banks were all boggy. They finally managed to churn it out but everywhere they'd been the road was cut to pieces with great deep tracks. Water was running everywhere and the entire area was completely slushy. They'd decided to return to the camp after running the Euclid's motor hot and out of water and cracking the cylinder head. Since there were no more 'wheels' left they thought they'd better stay put! About a week later I returned and found this awful mess with all my vehicles bogged everywhere and the Euclid buggered up. We had to pull the head off that and obtain another one. There wasn't much I could do about it because even though he was a bit hyperactive Robby was generally a really good worker. However, at that age he'd needed somebody there to supervise.

Eventually the water all dried up and then the countryside was completely transformed. There was massive growth of grass, shrubs and young trees to a height of about 3 or 4 feet. It was impossible to drive anywhere off the tracks. Even six months later masses of shrubs were growing everywhere and the dam was now a little lake and even the open-cut mine was still full of water.

Bushfires

The following year we didn't get any rain at all except for an odd shower. This resulted in huge areas of tall, dried up grass and shrubs. It was a tinderbox waiting for ignition and when the thunderstorms began the bushfires started. Since we were the only people out there with equipment suitable for fighting bushfires we were nearly driven mad with attending bushfires. We had a big water tanker mounted on 'Bertha' and we also used the Euclid and the fire plough. Nearly every afternoon a thunderstorm rumbled up and we'd see a lightning strike go down and sure enough, ten minutes later a big, black column of smoke billowed up and we'd have to race out with our crew to fight another fire.

We must have lost almost three months of working time fighting fires that year. We nearly lost 'Bertha' when a bushfire over-ran us. Her motor choked and stopped because the smoke, fumes and flames had deprived the air of oxygen. I had to clamber up on top of the truck until the fire passed. Even the little petrol motor up on top of the tank, with the jet hose for spraying fires, stopped because of the thick smoke. As soon as the fire had passed we started the motor again and hosed everything down, including the tyres which were alight in places.

Another time Lindsay and another chap were putting out a small fire about 20 miles away. In the distance Lindsay heard the roar of an approaching thunderstorm. One of those big strong winds rushed up ahead of the dry thunderstorm, then suddenly the fire was on him! It jumped across sandy river beds, a hundred metres wide, as if they didn't exist. The wind velocity was about 50 miles (80 km) per hour and the fire was racing along at about 20 to 25 miles (30 to 40 km) per hour.

They tried to outrun the fire in the Euclid, which only had a top speed of about 15 miles (24 km) an hour and the terrain where they were was rough going. Lindsay realised that he couldn't outrun it. He had the large, wide fire plough on behind, which could cut a track about 12 feet wide and could strip all the grass off. It was ideal for stopping most fires, but not one like this with a strong wind blowing. He judged the fire was only a couple of miles away and really racing towards him, so he quickly ploughed a circular area through the spinifex and pulled up in the centre. He elevated the 4-metre wide shovel about 15 feet in the air. They carried a 44-gallon drum of water in the shovel. The two men climbed up there and wet their clothing with the water in the drum and waited anxiously for the fire to blow past. Their hair was slightly singed but the fire raced past very quickly, being a grass fire. Immediately they put the siphon hose into the drum and extinguished the fire around the big tractor which was alight in several spots because tractors always have a lot of oil on them. The tyres were also scorching hot and just starting to burn. That was probably one of the narrowest escapes during that terrible year of fires.

During another of the chases we'd worked all day to put out a bushfire near Huckitta. All the station people had been fighting it, besides our crew using 'Bertha' with the fire plough behind. We finally had it under control and a night-watch crew were left on duty to make sure no falling trees set it off again or spot fires flared up again. After a while they reckoned everything should be okay so they returned to the station.

In the early hours of the morning the wind blew up and the fire was off again. The wind became quite strong and the fire started racing along at about 20 miles (32 km) an hour across the plains. Of course nobody had had much sleep the previous night, but during such times lack of sleep is a minor detail. Lindsay set off again to chase the fire, racing along one of the flats pulling the fire plough behind. At that point he was actually being chased by the fire. He came across a gutter which was hidden in the high grass and, in crashing through it, fractured the radiator. In the heat of the moment he didn't realise that he had lost all the water out of the motor which 'cooked' the head on that vehicle. Several of us were also out there in our assorted vehicles so I stopped to help Lindsay. We managed to keep 'Bertha' going by putting a siphon hose down from the big tank on the back to keep the water flowing into the radiator cap. It was only running on 4 or 5 cylinders then. Unfortunately that same day we lost the fire – it just went berserk! However, we managed to cut it off later on that night when the wind dropped.

Earlier that day I'd come out in a Holden ute bringing out some food and water to some of the men fighting the fires. I had to hurry back across one of the flats where the neck of the fire was racing across. I bounced over a gutter with the Holden and buggered up the radiator in that! The fan hit it but I couldn't even stop to have a look at it because the fire was approaching so rapidly. Eventually I veered out of the path of the fire, but I continued on, not realising I had lost the radiator water until I pulled up on a clear patch on the old Queensland road. By then the engine was really hot, so there were disabled vehicles everywhere. Since I'd left all the water with the other firefighters I had to wait there for help. I finally saw somebody coming along the road. It was Bruce Chalmers from Dnieper station. He was on his way home from visiting somewhere else but had come down the Plenty to help when he saw all the smoke from the fires. He gave me some water to put in the radiator which I blocked up and we made our way back to Marshall Bore and filled up with more water from there.

Things had just about settled down from the fires and I was bringing back a load of stores and fuel from town in the Holden one-tonner, loaded up with about a tonne and a half on the ute and a tonne and a half on the trailer. It was about 20 miles (32 km) from Huckitta, around ten o'clock at night and I was extremely tired. As I rounded a bend I ran into a mob of cattle 'shyacking' around in the middle of the road with the young ones racing about! I managed to skid almost to a stop, but even though I was probably only going about 5 miles (8 km) an hour I collected three young steers on the bonnet. One rolled right up on the bonnet and the other two just bumped the front. But, of course, the impact wrapped the radiator right around the fan and the motor!

I camped there that night and early next morning I pulled the remains of the bonnet and the radiator off, dumped them on the side of the road and blocked off the pipe to the water pump. I put water into the top hose to keep the motor full of water, but it took me about 3 hours to travel 20 miles (32 km). I'd go along for about 2 miles (3 km) before I'd have to wait until it cooled off again. Then I'd travel another couple of miles and stop again until eventually I reached Huckitta where I borrowed a vehicle and took the stores the remaining 20 miles to the mine. When I returned to the Holden I brought an old radiator with me and installed it. I eventually took the ute into town and traded it in on a Ford F100. I had really wanted another Holden but I would have had to wait about three months for one. My old Holden had been pretty well battered up from all the fire-fighting and fortnightly trips to town for stores and fuel.

On the very first trip out with the newly acquired F100, blow me down, I ran into rain which meant a lot of detours and I became bogged several times. The last bog was about 3 miles from Molyhil camp so I walked in for the last leg of the journey. I returned the next day with a truck to get the Ford. Even the truck was bogged on the way out there but I managed to bring the stores in. A couple of days later I was able to get the ute out.

Another Escapade by Robby and friends

About three trips later while in town, I was delayed and couldn't take the stores out so I sent

Robby Warner, who was in town with me, back to Molyhil with the truck. Everything was fine on the trip out. He did a good job and arrived there safely, but a couple of other white fellows out there wanted to go from Molyhil to Bonya. They helped themselves to some of the grog among the stores and decided, "Yeah, let's go to Bonya." A couple of the young bucks knew some of the women over at Bonya, so they all set off in the F100 that night without unloading the ute first because that was too much trouble.

They came to a big washout at Oorabra Creek where there was a detour. Since it was night and they were half full they didn't see the detour until it was too late. The washout was a sheer drop of about 5 feet. They just crunched down, with the nose of the F100 going straight down into the hole and striking the heap of sand, which bent the chassis a little and buckled up the front of it. Robby still had his motorbike on the back of the truck (he hadn't even taken that off!) so he rode that back to Molyhil, got the Euclid and some heavy cables and lifted the F100 out of the big hole. By that time they were all sober and sorry! They gave up the idea of going to Bonya! It was next morning by then, anyway. Such were some of my trials and tribulations! Never a dull moment!

The next thing sent to plague us really was a plague! After the big Wet many of the aborigines around the countryside and a few of the whites (who actually brought it up to Alice Springs) were stricken with a plague of scabies. We had to go around and inspect the camps for evidence of scabies. If we found it clothing had to be burned, the people had to be scrubbed and a gooey ointment applied. After about twelve months the scabies scourge which had spread to most parts of 'The Centre', was finally eradicated. There were also outbreaks of syphilis and gonorrhoea through most of the camps in town – whites and blacks alike – hundreds of cases of it. We were cast into the role of being 'Doctors', giving injections and so forth since all the people affected couldn't be carted into town. The Flying Doctor would come out and check everybody but we had to do follow-up injections on the spot.

Between managing the mining, being a transport operator, motor mechanic, firefighter and so forth it was one continual saga. I suppose it is no wonder my 'ticker' gave way again. Eventually I decided the best thing to do was to sell out and retire. The mine was well enough developed to be a good proposition for a potential buyer.

Kurt feeding the animals from the kitchen of Mulga Express Mark IV. (Photo courtesy Carmel Sears, 'Centralian Advocate')

PART VI
RETIREMENT (1979 onwards)

Chapter 44
'Mulga Express' Mark IV and My Second Wood-gas Producer

In 1976 I bought a 1972 Dodge Coronet Station Wagon, left-hand drive, from an American serviceman who had been stationed at Pine Gap, Northern Territory, and was returning to U.S.A. I modified it by strengthening the suspension and sub-chassis, replaced the fourteen-inch wheels with sixteen-inch light truck wheels and tyres, giving it better clearance for outback 4WD tracks. I designed a screened-in sleeping tent on a special folding rack mounted on top of the vehicle, which pulls down with a small annexe for dressing. A step-ladder hooks on the side to climb up into the double-bed and I added an extra double-bed extension over the rear section of the vehicle with some steps welded onto the left rear corner. The rear bed folds out beneath an extra self-supporting annexe over the entire kitchen area, netted in right around. There is ample floor space for another swag or two on the ground. Green outdoor carpet is laid down in sections, pleasant to walk on and saves a lot of bindi-eyes or dirt getting into the beds.

The kitchen cupboard is 2.5 metres long and slides out on a counter-lever. When it is pulled out the door, which closes against the cupboard, folds down and acts as a table. Inside are six sliding drawers for storing groceries and three open compartments. When the door-cum-table is up it covers all except the two open compartments nearest to the front seat, where things which need easy access while travelling can be stored. Attached to the front end of the cabinet a two-way CB radio is mounted, with access from the driver's seat when the cupboard is pushed in, or from the back of the car when the cabinet is slid out. The rear part of the sliding cabinet contains a refrigerator which can be easily accessed through the back window. The fridge is a three-way type which is mainly run from a 12-kilogram gas bottle. The gas is used for both the stove and the refrigerator and generally lasts for about three weeks, which is quite economical. The two-burner gas stove is built into the kitchen cabinet, next to the refrigerator.

I also carry three fluorescent lights which operate from the 12-volt heavy-duty car battery. I have fitted a special 240-volt 3,000 watt alternator which can be used for my mechanical tools such as electric drill, angle grinder and arc welding. I also have oxy-welding gear and a full set of general garage tools, plus bits and pieces for emergency use. All this weighs about 75 kilograms, but the vehicle has been worth its weight in gold in the real Outback, both for me and other travellers in trouble I have met up with along the way.

Building the Second Wood-Gas Producer

In 1987 I built and installed my second wood-gas producer after recuperating from heart surgery in December, 1986. The first one I built in 1940 has been described in Chapter 23. The gas producer is mounted on a frame which swings by a hinge out and away from the back of the Dodge. This means all the kitchen gear can still slide out. This wood-gas producer has functioned extremely well and has now travelled many thousands of kilometres since it was installed. It can be disconnected and left off the vehicle if I wish to travel around town. When travelling with the gas-producer installed I can run totally on wood-gas, or petrol, or a combination of both. It is a fair bit of weight to carry around, but with the right vehicle that is not a problem. It certainly makes exploring around in the bush or prospecting a pleasure.

My camping vehicle, 'Mulga Express' Mark IV (Photo courtesy Carmel Sears, 'Centralian Advocate')

Since I installed my wood-gas producer in 1987 I have carried a chain-saw, which I call my "bowser." There is also my prospecting gear – a pick and axe, a couple of gold dishes and a sieve. There are two small jacks and a little hand-winch plus a high-lift bog jack and tow ropes of about 30 metres each plus a tow chain about 3 metres long, all of which have been useful when I've met up with people who have become bogged or needing towing to the nearest garage. There is a spare tyre and wheel, a coil of wire, two tyre pumps (one electric and one spark plug pump), two spare tubes, patching gear, a pair of jumper leads, a siphon hose, and a can of petrol. The fuel tank holds 160 litres which is excellent for long bush trips. I store 200 litres of water in tractor tubes mounted around and underneath the front seat and floor boards, fitting into all the nooks and crannies. There is room in the kitchen cabinet for enough general stores to last about four or five weeks. I also carry some fishing gear. There isn't a toilet, but there is a 'holey' chair which serves the purpose rather well and provides excellent views! A shower can be rigged up inside the annexe or under a tree. As you can see, my 'Yank Tank' has been altered quite a bit but it has really served me well so far and enables me to go 'roughing it' in comfort!

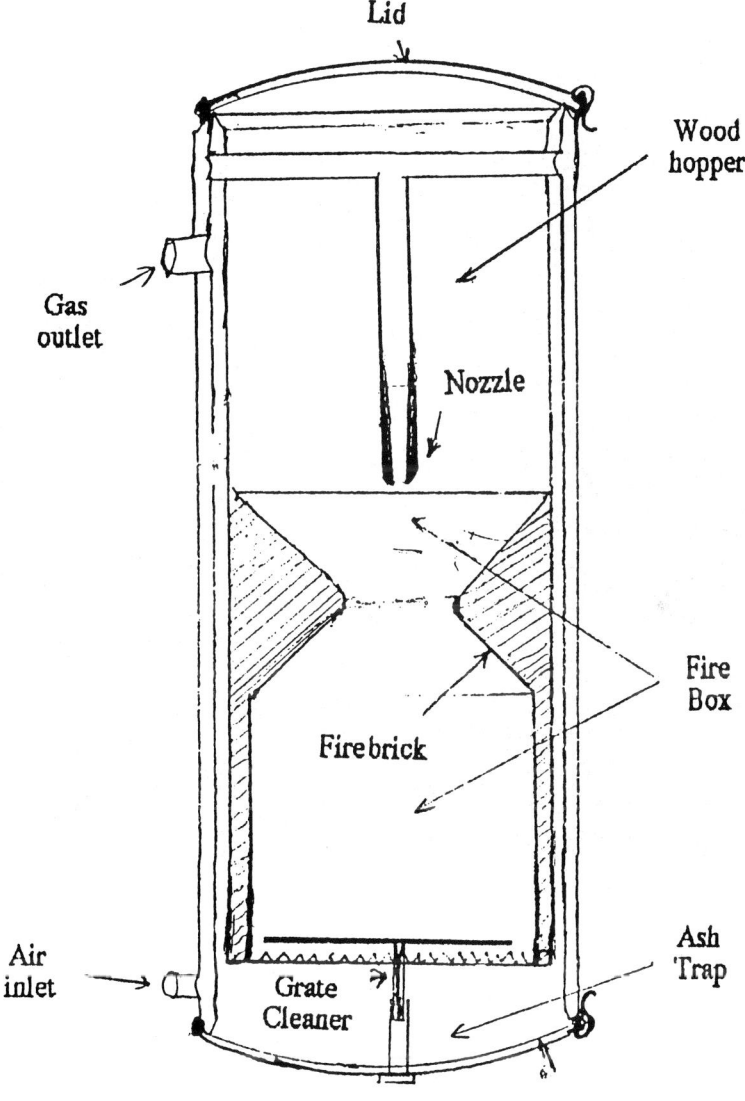

A sketch plan of the Wood-Gas Producer.
(Full plan available from the author).

Chapter 45
Keeping on Keeping on

After I retired from mining and transporting operations in 1980 I purchased a small hobby farm in a lovely, picturesque spot near Yankalilla, South Australia. I bought a tractor and repaired a chaff cutter with the intention of selling chaff locally. I built up the sheds and renovated the old house which had been built around 1850. However, we sold it in 1984, as Elsie didn't particularly like living there and we were going our separate ways by then.

I had broken my ankle quite badly in 1981 which forced me to slow down considerably for about two years. That accident happened back at Alice Springs when I was showing an inexperienced driver from Plenty River Mines how to correctly load his semi-trailer. He was ready to leave with about 4 tonnes of bits and pieces scattered over the trailer. He said he couldn't fit any more on. To his surprise I managed to load another 8 tonnes of urgently needed equipment on board. As I was finishing securing the load I caught my foot on a piece of wire on top of the load, throwing me off balance and forcing me to take a backwards leap off the side of the truck. I landed on some wooden planks about 2 metres below, smashing my right ankle, which was set by a doctor in Alice Springs. After about a year my ankle still hadn't healed very well so I went to a specialist in Adelaide, but little could be done to improve it by then.

My life has been interesting and full of challenges. I have met many wonderful and kind-hearted people as well as my fair share of rogues, liars and cheats! My mind still ticks over with ideas for things to make and do, and the bush country of Outback Australia continues to call for me to come and explore some more!

In 1984, I went north on a trip to Ayers Rock with friends, Hugh and Marge Holloway and Dudley and Helene Burns, taking two vehicles. We had 'Noddy', a little red International ute which I drove, while Dudley and Hugh took turns driving my 1972 Dodge station wagon. Near Kingoonya the Dodge hit a rock causing an oil leak in the transmission. Hugh and Dudley took the ute to go and get some more transmission oil. They first called in at Mount Eba but there was no oil available there so they had to go on to Coober Pedy. On their return at night they missed the turnoff to Mount Eba where we were waiting and travelled quite a way down to the bitumen highway before they realized they'd missed it. They eventually arrived back at our camp around 3 a.m., freezing cold and not very happy! We continued on our way to Ayers Rock, but after we left there the Dodge's transmission really malfunctioned due to the loss of oil and we had to tow it into Alice Springs. I took a transmission from a wreck and installed that into the Dodge which I was to really regret later during a trip to the Bungle Bungles! However, we had some enjoyable times on that trip.

Between 1983 and the present time I have travelled with friends to many places in the beautiful North country of Australia, including the Bungle Bungles and Kimberleys in Western Australia, the Top End, the bottom end of the Gulf of Carpentaria, the Simpson Desert, the Western, Central and Coastal areas of Queensland and many of my favourite haunts in 'The Red Centre'.

One of my greatest enjoyments is in sharing with my friends the wonderful scenery of the Outback, especially in the more remote areas where few people visit. The colour section shows some of those places.

After my heart played up again on a trip in 1986 I had triple-by-pass heart surgery in December, 1986 and have recuperated really well since then and am able to lead a pretty active, normal life. During my recuperation I commenced tape-recording my memories, which was the beginning of this book. I also tried my hand at oil painting 'from scratch' and have painted some views of my beloved Territory, including one from memory of my childhood home of Deep Well Station.

Many years spent beneath the bright northern sun, plus many years of using arc welding equipment, possibly contributed to the growth of cataracts on both of my eyes. I had the cataracts removed and lenses inserted, the first one in 1988 and the second one early in 1989, both of which were successful, with my sight in both eyes now being excellent. I was advised to avoid the excessive heat of summer up in Alice Springs so I usually spend about November to April in Adelaide or somewhere south and from May to October travelling in the northern country of Australia or spending time in Alice Springs with family and friends and showing others some of my favourite haunts in 'The Red Centre.'

Elsie and I have been separated for the past few years and the rest of my family are living both in the North and South. Lindsay and Joan still live at "Baikal" near Bonya and the Jervois Range where Lindsay still does a bit of prospecting. He and Joan run a store which supplies food and general provisions to the aboriginal community at Bonya and to nearby stations. David and Bevlee and family live in Alice Springs where David owns and manages The Equipment Centre, a hardware/tool business. Peter and Christine live in the southern area of Adelaide where Peter works as an inventor and electronics technician for the Laserex Company. Paula is living nearby in Adelaide with her husband, Timothy Hall, their little son and baby daughter. Young Kurt and Vicky now own our old home at Mount Charles in the Alice, with its lovely view of the MacDonnell Ranges. They have recently taken up a service station business opportunity in Yulara and will give that a go for a while with their young family. At the time of writing in 1992, I have 5 children, 16 grandchildren and 9 great-grandchildren.

Besides travelling I have been working on family history and writing this story, organizing photos and historical aspects of road transport. Both my first road trains, 'Bertha' and 'Wog' are being restored and are now residing at the Ghan Preservation Society's Museum in Alice Springs. It is planned for them both to be housed in the proposed Road Transport Hall of Fame in the Alice.

I have led a very active, adventurous life with many challenges and problems to overcome, but it has been rewarding and good and some of my dreams have been achieved. Looking back over the years I don't think I'd want to change much. My life is now a part of the history and development of the Territory and, even though it may only be a small part, I am proud to have contributed something in that regard.

Appendix A: Family Tree Chart of Kurt Gerhardt Johannsen

Gerhardt Andreas Johannsen
b. 14.11.1876 Denmark
d. 4.4.1951 Alice Springs, N.T.

Marie Ottilie Hoffman
b. 15.7.1882 Barossa Valley, S.A.
d. 26.12.1959 Alice Springs, N.T.

m.1905 Barossa Valley, S.Australia

Children:

1. Elsa Margaret Johannsen
   b.21.6.1906 Barossa Valley S.Australia
   d.11.12.1986 Alice Springs, N.T.
   m.Wilhelm Petrick

2. Curt Johannsen
   b. May 1910 Barossa Valley, S.Australia
   d. Nov.1910 Barossa Valley, S.Australia

3. Gertrude Ottilie Johannsen
   b.28.8.1912 Deep Well, N.T.
   m. Alan M. Hayes

4. Kurt Gerhardt Johannsen
   b.11.1.1915 Deep Well, N.T.
   m1. Kathleen Rowell 1938. div.1948

   Children:
   (1) Lindsay Andrew Johannsen
       b.18.7.1939, m.Joan Fletcher
   (2) David William Johannsen
       b.6.4.1943, m.Bevlee Olive
   (3) Peter Kurt Johannsen
       b.3.5.1945, m.Christine Jelly

   m2. Daphne Avis Hillam 1950. div. 1958
   m3. Elsie Dixon Collins 1958

   Children:
   (4) Paula Francesca Johannsen
       b.4.11.1959, m.Timothy Hall
   (5) Kurt Damien Johannsen
       b.28.11.1964, m.Vicki Lane

5. Mona Dora Johannsen
   b.27.10.1923 Hermannsburg, N.T.
   m.Desmond Byrnes

6. Randall Werner Johannsen
   b.10.8.1925 Deep Well, N.T.
   m. Ruth Cummings
   d.13.11.1997 Adelaide, S.A.

7. Myrtle Edna Johannsen
   b.22.11.1926
   m.Albert Noske

Appendix   247

Appendix B: Map of some of the Pastoral Leases in Central Australia, 1914

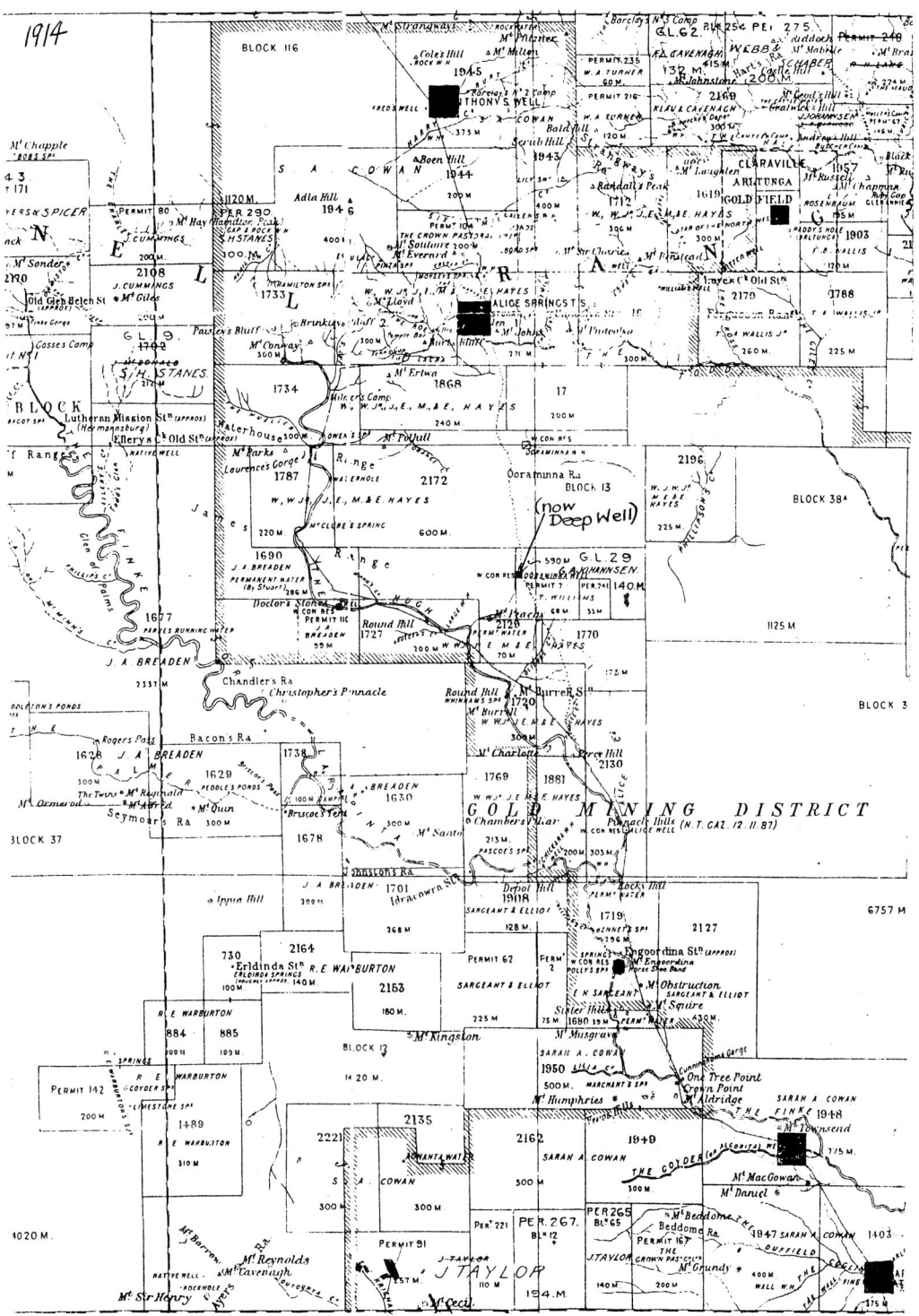

Appendix C: Map Showing My Transport Operations in the Northern Territory

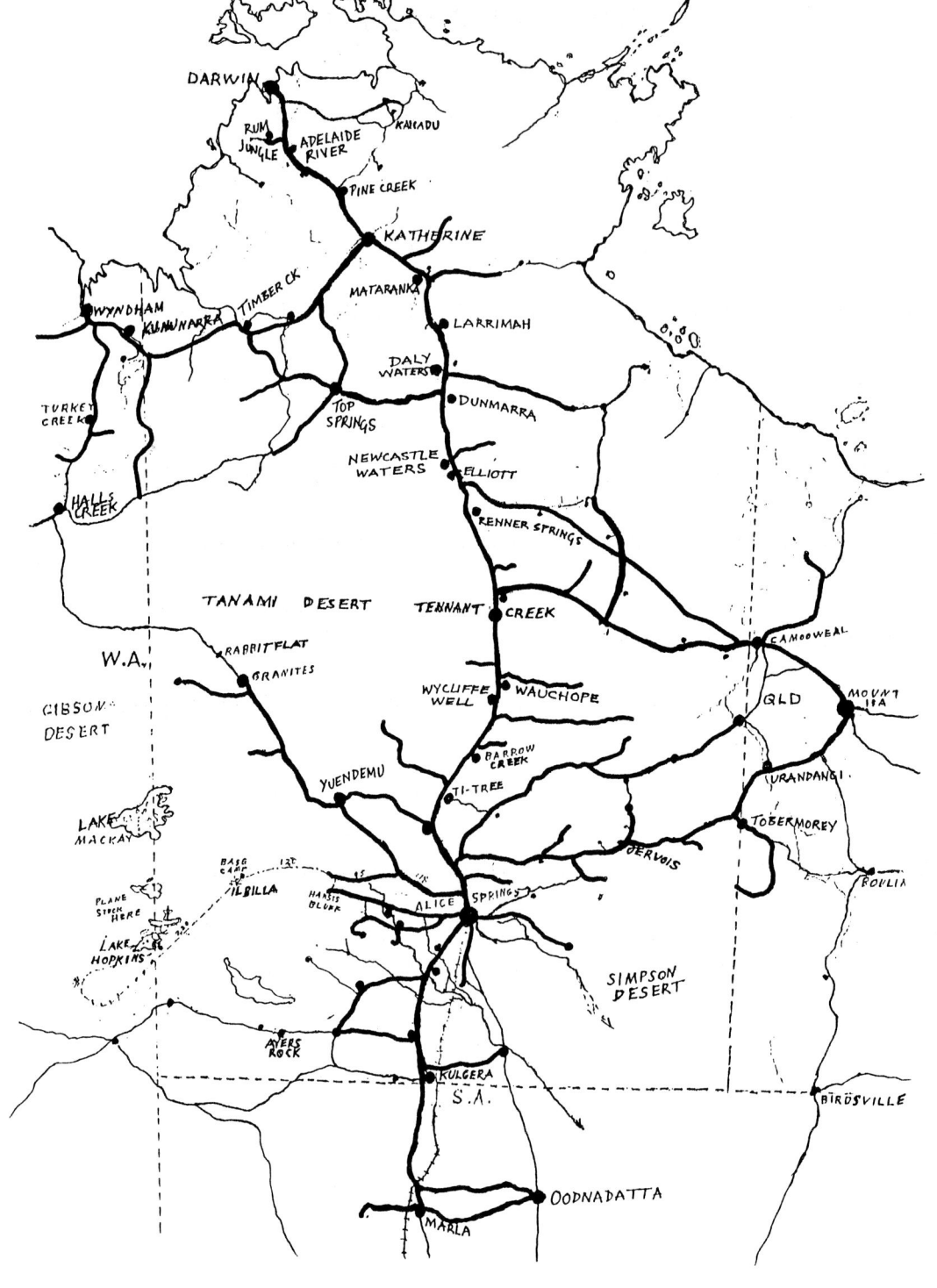

FURTHER READING

Australian Geographic magazine, October, 1991, article on Ryan's Well.

Grant, Arch, *Camel Train and Aeroplane* (The story of Skipper Partridge), Rigby, 1981

Keneally, Thomas, *Outback*, Hodder & Stoughton, 1983.

Isaacs, Jennifer, *Pioneer Women of the Bush and Outback*, Landsdowne Press, 1990.

Maddock, John, *A History of Road Trains in the Northern Territory 1934-1988*, Kangaroo Press, 1988.

Petrick, Jose, *The History of Alice Springs Through Street Names, Graphic Services, 1989.*

Pfitzner, S.G. and G.E., *Arbor Hoffman II*, Lutheran Publishing House, 1986.

Pioneering History of the Johannsen Family, compiled from family notes by Jose Petrick, Eugene Pfitzner, Trudy Hayes and Kurt Johannsen (unpublished).

Plowman R. Bruce, The Man from Oodnadatta, (collected and abridged by Jean Whitla), Shoestring Press, 1992.

New Outlook, House Journal of the Colonial Mutual Life, August, 1980.

Wauchope, Alan, "Seeking the Golden Reef" *Wide World Magazine*, June 1953.

Photographs, headlines and information contained in newspaper reports from *The Advertiser*, S.A., 1950 and the *Centralian Advocate*, Alice Springs, 1950, 1974, 1987.

INDEX

A.I.M., 15, 41, 46, 60, 69, 138
Aborigines, 9, 15-16, 27, 29, Ch. 5, Ch 6, p. 51-2, 66, 80, Ch. 12, p.93, 120-1, Ch. 35, p.207 216-7, Ch. 40, Ch. 42
Accidents,
  Dad, 7, 69
  Kurt, p. 10 Ch.24, Ch.27, Ch. 30 Ch. 33, Ch 34, Ch.35, Ch. 42
Adams, Steve, 27
Adamson, Doug, 78
Adelaide 2, 25, 29, 37, 41-3, 47, 72, 95-8, 124, 138, 153, 163, 166-7, 207
Adelaide House, 15, 37, 138
Aebart, Dr. 75-6
Aeroplanes, 21, 55, Ch.30, p. 165 Ch.35
Afghans, 4, 7, 15, 21, 28, 49
Aileron, 72
Albrecht, Pastor Freidrick, 33, 60
Alcoota, 63-4, 89
Alice Springs, (mentioned throughout book)
Alice Well, 8, 30, 43
Algebuckina, 28
Allied Works Council, 126 134-8
Althouse, Jack, 81-2, 116-7
Ambalindum, 63, 65-6
Ammaroo, 66, 218-9
Anderson, Sandy, 210
Angaston, 2
Angel, Mr, 153
Angle Pole Waterhole, 27
Anthony Lagoon, 73, 150
Anthropologists, Ch.12
Archer, Flt.Lt.J., 162
Argadargada, 76
Arltunga, 8, 23, 63-6, 73-4, 77
Armstrong, Cecil, 135-6
Arthur River, 200
Attack Creek, 103
Aulgana Mica Mine, Ch.16
Aunger, Murray, 35
Australian Solar Ponds, 189
Avon Downs, 135

Ayers Rock, 91-4, 185, 244
Baikal, Ch. 42
Bails, Arthur & Mrs, 19
Baker, Mr, 58-60
Baldissera, Joe, Ch.40, p. 218
Baldock, Murray, 115,
Ballarat, 97
Banka Banka, 72, 80
Barcaldine, 131
Barkly Hwy, 74, 154
Barossa Valley, 2, 25, 46
Barrow Creek, 72, 103, 115, 143
Basedow, Dr. Herbert, 25
Bathan brothers, 81
Becker, Toby, 166-9
Bell Bird lease, 210-11
Bell, Neville, 145-7
Ben, Emily, 235
Bennett & Fisher, 50
'Big Foot', 186-7
'Billy Hughes' mine, 101
Bird, Jimmy, 89
Birdsville, 189
Birdum, 70-1, 103, 107, 113-18, 122
Birtles, Sir Francis, 21-2, 55
Bitter Springs Gorge, 76
'Bitzers', 91-2, Ch.16, Ch.18
Blackall, 131
'Black Eagle' Mine, 67
Blitz, Mr, 211
Bloods Creek, 19, 30
Bob-the-Dog, 201
Boer War, 2
Bond, Bert, 73
Bonds Tours, 73
Bonning, Jack & Mrs, 80
Bonny Well, 103
Bonya Hills, Ch.31, Ch.42,
Bordertown, 96
Borroloola, 164
Bowes, Jim, 153
Brandt, Gus, 77, 107-110
Breaden's Dam, 9
Branson, Syd, 105
Brisbane, 126, Ch.23, Ch.24
Brocks Creek, 122, 155
B.H.P.Ltd., 75, 172
Brookes Soak, 72, 120

Brookes Soak Massacre, 66, 72,
Broome, 248-9
Brown, Mick, 212
Bruce, Bluey, 225
Building houses
  Deep Well, Ch.1
  Alice Springs, Ch.7, p. 157
  Flagstaff Hill, 222
  Mount Charles, 222
  Baikal, 225
  Undoolya Rd., 159
Bullwaddy Boys, 81-2
Bungle Bungles, 244
Bureau of Mineral Resources, 168
Burma, 102
Burrell, Mount, 9
Burns, Dudley & Helene, 244
Burton, Miss, 53-4
Bushy Park, 89
Buttfield, Mount, Ch. 35
Byrnes, Des, 141

Caltex Company, 153
Camooweal, 120, 199-200
Capitol Picture Theatre, 127
Carroll, Bill, 106
'Caruso' mine, 101
Cattle transport, Ch.28, Ch.37
Cattle Transport Assoc.199, 204
Calder, Sam, 188
Cavanagh, Fred & Mrs, 65
Cawood, Stan, 57
Central Mount Stuart, 115
Central Pacific Minerals, 224
Chalmers, Bruce, 238
Chalmers, Mac, 74
Charles, Mount, 222
Charlotte Waters, 27, 30
Christmas Dam, 9, 16
Ciccone, Patsy, 68
Claraville, 63, 65
Clare, 30, 99
Cleland, Professor, 84
Cloncurry, 120, 131, 135
Clough, Hughie, 60
Colback, Bill, 51
Collins, Elsie, 213
Colson, Bill, 120

Commonwealth Dev.Bank, 144-5, 214
Commonwealth Rlwys, 158-9
Coniston, 72, 120
Connair, 52, 207
Connors Well, 143
Conway, Alec, 116-7, 122-4
Coomalie Creek, 169
Coppock, Jim & Lance, 205
Corroboree, 15
Costello, Joe, 25, 32, 36-7
    Alice, 9
Coxen, Bob, 185
Crashes, plane, Ch.30
Crowther, Norman, 67
Cummings, 151
Cummings, John, 221

Dajarra, 117
Dale, Postmaster 114-5, 124,
Daly Waters, 72, 93, 118, 123
Davey, Jim, 27
Davis, Pat, 161, 174, 183
Davis, Professor, 84
Darwin, 72, 81, 106, 113, 122-3, 145-9, Ch. 29 p.166
Deakin, Charlie, 103-4
Deep Well, Chapters 1-6, & p.57
Dekane, Bob, 137
Delny Station, 63, 89
Denny, Tom, 157
D.C.A., 163, Ch. 35
Department of Works, 105
Depression, 72, 97-99, 104
Depot Sandhills, 30
Devils Marbles, 103
Dingoes, 12, 68-9
'Disputed' mine, 101
Disposals Commission, 153-4
Dixon, Ted, 142-3, 150
Dnicpcr Station, 238
Docker River, 93, 187
Doreen, Mount, 72, 243
Dreamingtime, 45
Drought, Ch.5, Ch.6, Ch.38 Ch.43
Drum Case, Ch.29

Duchess Mine, 131, 135
Dunmarra, 72, 113, 118, 157-8
Dunstone, S., 162

Dutton Party, 35

East End Market, 98
Eba, Mount, 161-3
Ebenezer, 2
Education, Ch.2, Ch.5, Ch.7
El Dorado Mine, 82, 96, 108, 123
Elkedra Station, 141
Elliott, 169
Elliott, Mr & Mrs Gus 25-6
Emerson, Mr, 163
Emily Plains, 15
English, Jimmy, Ch.35
Entire River, 235
Erldunda, 139
Eskimoes, 29

'Faith in Australia', 54-5
Fanning, Jerry, 155-6
Favaro, 157
Fidler, Frank, 170-1
Finke River, 25-6, 30
Fires, Ch.43
Fitzpatrick, Bryan, Ch. 41
Fitzpatrick (Fitzy), 104-5
Fitzgerald, Sergeant, 79-80
Fleming, Stan 58
Floods, 26, 57, Ch. 19, Ch.43
Flynn, John, 37
Fox, Billy, 60
Foy, Vic, 91-94, 130
Frederiksen, David, 140, 189
French, Gene, 23, 77

'Gander' Mine, 70
Garden Station, 66
Garland, Len, 67
Gates, Gitta & Hadley Ch.24,
Gearbox Hill, 201-2
Georgina Basin, 219
General Mica Supplies, 100
General Motors Agency, 141
German Joe, 67
'Ghan' train, 28, 60
Ghan Preservation Soc. 245
Glass, George, 70
Golder, Claude, 23, 77, 109
Golder, Tim, 77
Goldfield Hotel, 79
Graham, Clive, 155
Grand Coffee Palace, 29

Granite Downs, 27,
Granites, The, 67, 72
Greatorex, Kath, iii, 164
Gregory, Bob, 98-9
Guinea Airways, 72, 123
Gunster, Maria Dorothea, 2

Haasts Bluff, 84, Ch. 35
Hall, Eddie, 151
Hall, Punch & Judy, 127
Hall, Tim & Paula, 245
Halls, Bro, 80
Hanlon, Tom, 205-6
Hanlon's Extended Mine, 168
Hamilton Downs, 84, 174
Hanson Swamp, 115
Hardy, James Co. 33
Harris, Ken & Claire, 95-6
Harry, 9, 44
Harts Ranges, 46, 51, 64, 67, 70 100-2, 125, 136, 237-8
Hatches Creek, 165, 171-5
Hayes, Alan, 4
Hayes, Bill, 4, 27
Hayes, Strat, 27
Hayes, Ted, 27, 31, 44
Hayes, Alan and Trudy, 3
Healy, Noel, 118, 157
Heart attacks, 220, 231, 245
Heavitree Gap, 15, 52
Helen Springs, 80-1, 103-4, 116
Hermannsburg, 2, 3, Ch.5, 37, 60-1, 85, 103
Hersey, Arthur, 77
Hicks, Jack, 82, 96-7
Hillam, Daphne, 164, Ch. 33-8
Hodgman, Jim, 154-7, 165
Hoffmann, Karl J., 2
Hoffmann, Maria Dorothea, 2
Hoffmann, Ottilie, 2
Hoffmann, Gertie, 96-7
Holloway, Hugh, Marge, 244
Holtz, Wallaby, 80
Homburg, Marcy, 61`
Home of Bullion Copper Mine, 166-8
Hong Kong, 220
Hopkins, Lake, Ch. 35
Horseshoe Bend, 25, 30, 109
Horwood, Edgar, 23, 77
Huckitta, 74-5, 235
Hugh River, 30

Hughes, Jack, 60, 68
Hunter, Nick, 182

Ilbilla, Ch.35
Immanuel College, 97
Indiana Track, 217
Isaacs, Jennifer, 3
Irvine, Sam, 23, 41, 63, 77,
  107, 111, 113
Isa, Mount, 73-4, 107, 131-5,
  150
Italians, 64, 100-2, 137

James, Mrs, 209-10
Japan, 220
Jasper Gorge, 202
Jay Creek, 21
Jervois Mine, 107
  Ch.24, Ch.33, Ch.38, Ch.39 Ch.42
Jervois Range, 75-6, 165, Ch.38
Jinka Springs, 74-6, 205
Johannsen family
  Curt, 21
  Daphne, Chs. 33-38
  Elsie, 215, 220, 222, 244-5
  David & Bevlee, 161, 164, 245
  Elsa, 2, 9-15, 25-6, Ch.6,
    Ch.7, p. 171
  Gerhard, Chs.1-9, Ch.8,
    pp 123, 128, 137-8
  Kurt Jr & Vicky, 213, 222, 245
  Lindsay and Joan, 128, 131-6,
    161, 164, 169, 200, 207, 245
    Ch. 41
  Matthias, 2
  Mona, 32, 39-40, 71, 75,
    137-8, 141
  Myrtle, 36, 40, 137-8, 157
  Nickolas, 2
  Ottilie, Chs.1-9, p.128, 137-8,
    172
  Paula, 213, 215, 222, 245
  Peter & Christine, 161, 164, 245
  Randle, 36, 40, Ch.33, 222
  Trudy, Chs.1-9, p. 73, 128
Johnny Barboon, 9, 44-6
Johns, Paul, 71-2, 96-7
Jones, Jock and Mick, 81
Jones, Norman and Mrs, 51

Kalkarinji, 199

Kangaroo Well, 44-5
Kaporilja Springs, 33
Katherine, 113, 123, 149, 203
Kenna, Snow, 70, 129
Kerin, Mick, 25-6, 36
Kerr, Alec, 63, 89-90
Khan, Fred & Hussein, 8, 77
Kidman, 74
Kilgariff, Joe, 78-9
Kimberleys, 244
King, Virgil, 145
Kingoonya, 24, 95
Kingsford-Smith, Chas. 54
Kingston, S.A., 98
Kleinig, Martin, 40
Koerner, Carl, 143, 157, 161
  182, 185
Kooraroong, 44-5
Kramer Ernest & family 19-20
  46
Kramer, Colin, 20
Kramer, Mary, 19-20
Kunoth, Henry, 157, 200, 202
Kunoth, Sonny, 36
Kunoth, Trott, 36
Kunoth, Ted, 157

Lackman, Jimmy, 51, 60, 127-8
Lady, 9, 44
Lake Hopkins, 177-84
Lake Nash, 74
Lake Woods, 81, 117
Larrimah, 159
Larkin, Bob, 138
Lasseter, Harold Bell, 71
Lasseter's country, 93, Ch.35
Liddle, William (Billy), 8
Liebig, Mount, Ch.12, Ch.35
Light Pass, 29
Lime, cement & pit, 5-6
Lincoln Bomber, Ch.35
Line parties, 22
Little, Daphne,
Llewellyn, Bill, 157-8
Longreach, 131
Lutheran Church, 32
Lyons, Leo, 155

McCartneys Transport, 96-7
McClaren Creek, 105
McConville, Harry, 96

McCoy, Billy, 23, 77, Ch.35
McDonald, Al, 79, 129-30
McDonald, Len, 53, 57
McDonald, Mr Len, 57-9
Macdonald Downs, 63, 74, 76
MacDonnell Ranges, 26, 125
McEwen, John, 189
McGraths, 153
McIntyre, 42-3
McKay, Ross, 183-5
McMahon's, 79
McNab, Sandy, 166-7

Mackay, Constable John (Jack), 43
Madrill, Bill & Maude, 74
Maggie, 9, 44
Mail runs, 24, Ch.9, Ch.10,
  Ch.18, Ch.19
Malcolm Moores, 130-6
Mallala, 162
Maloney, Jim, 82, 142
Manners Creek, 208
'Man-powered', Ch.24
Mansell, Darkie, Ch.35
Marinji Track, 169
Marquar Station, 212
Marree, 28
Martin, Cock, 80
Martin, Doreen, 53, 60
Martin, Herb, 19-20
Mars Machine Tools, 133-
Marsh, Vic, 133
Marshall River, 235
Mary, Mount, 217
Maryvale Station, 4, 9, 27, 30
Maskell, Jack & Edna, 207-8
Mataranka, 113
Mawby, Sir Maurice, 168
May, Jack, 230
Melbourne, 57, 95-98, 100, 130,
  153, 168, 214
Melrose S.A, 30
Meyers, Mrs, 57
Mica Mines, Ch.16, Ch.25
Menz, Mrs, 61
Middleton Ponds, 91
Miller, Damien, 227
Miller, Eric, 82-3
Milnes, Musty, 36-7, 51
Milners Swamp, 116-8
Mines Dept., 71, 211, 225-6

Index

Mobil Oil, 153
Molyhil Creek, 234
Molyhil Mine, Ch.42, Ch.43
Moores, Malcolm, 130-6
Morley, John, 212
Morton, Nugget, 66
Mount Burrell, 9
Mount Buttfield, Ch.35
Mount Charles, 222
Mount Doreen, 72, 243
Mount Eba, 161-3
Mount Gambier, 98
Mount Isa, 73-4, 107, 119, 131, 152
    208-9, 215
Mount Liebig, Ch.12, Ch.35
Mount Lyell-Brown, Ch.35
Mount Mary, 217
Mount Palmer, 100-2, 137-8
Mount Riddock, 63-4, 70, 205
Mount Swan, 63, 65, 90, 102, 165
Moyer, Alan, 122, 131
Muckaty Bore,
Mud Tank, 205
Mudge, Graeme
Murray Downs, 150
Murray, George, 66

Narwietooma, 183
National Aust.Bank, 52
Nash, Lake, 74
Newcastle Waters, 8, 73-4, 77-8,
    81-2, 103, 114-8
New Well, 39, 42, 44-5
Nicker, Ben, 91, 93, 109, 120
Nicker, Claude, 74-5
Nichols, Arthur, 105
Nichols, George, 77-8, 105
Nine Mile Mine, 68
Ninety Mile Desert, 98
Nive River, 132
Noble, Jack, 78
Nobles Nob Gold Mine, 78,
    166
Noornee, 27, 127
Northern Drillers & Co.
    168, 206
Northern Motors, 141
Noske, Bert & Myrtle, 157,
    165, 200
Noumea, 220

Ochtman, Mr, 25
Old Crown Station, 26, 30, 109
Old Huckitta, 74
Old Timers' Home, 9
Old Timer's Museum, 10
Olgas, 93
Oodnadatta, 2, 4, 18, 19, 23-8
    30-1, 41-43, 47, 49, 77-8,
    99, 103, 109
Oorabra, 74, 205, 239
Owen, L.M. 128
O'Malley, Pat, 122-3
Owen Springs, 4, 91

Painters Spring, 87
Palm Valley, 35, 61, 73
Palmer, Mount, 100-2, 137-8
Parafield Airport, 161
Partridge, Kingsley, Rev. 19, 97-8
Pearce, Bill, 124
Pech, Ruth, 60-1
Peko Mines, 140
Pell Airstrip, 155
Perkins, Alice, 9, 21, 41
Perkins, Burk, 9, 39
Perkins, Emily, 9, 21
Perry, Con, 82, 174-5
Petermann Ranges, 71, 185
Petrick, Bill, 46, 51, 56, 64,
    90, 102, 165, 171
Petrick, Elsa, 51, 90, 102,
    165, 171
Petrick, Josie, 3, 9
Petrick, Martyn, 41, 47, 165
Petrocarb Exploration, Ch.41-42
Pettit, Ted, 150, 157
Pfitzner, Eugene, 3
Phillip Creek, 82, 116
Pick, 'Taffy', 105
Picton Springs, 74-5, 206
Pine Creek, 113, 168
Pine Hill, 36, 72
Pines, The, 95
Pinnacles, The, 206-7
Pinnacle Well, 125
Pioneer Theatre, 129
Pizzinato, Peter, 101
Plaza Private Hotel, 30
Plenty Hwy, 74
Plenty River, 74, 205, 235
Plenty River Mining, 224

Plowman, Bruce Rev., 1
Poseiden, 224
Powell Creek, 72, 80, 103
    113
Price, Bluey, 220
Price family, 21
Prince, Jimmy, Ch.28, Ch.35

Quartz Hill, 74
Quorn, 30, 99

Race, motorbike, 62
Radios, 37, 57-8
Railway to Alice, 49-50
Ramsey, 225-6
Rapson, Mr., 133
Rawlinson Ranges, 176
Red Tank, 205
Reiff, Norman, 235
Reilly, Billy, 141
Reilly, Molly, 137, 141
'RESO' party, 14, 35, 47
Riddock, Mount, 67, 70, 207
Riedel, Pastor John, 35
Riley, Paddy, 60
Ritchie, Peter, 157, 200-2
Road trains (see Contents)
Road Transport Hall of Fame, 245
Rodericks Bore, 118
Rodericks Swamp, 124
Rogers, Dick, 204
Roma, Queensland, 132
Rose, Lionel, 204
Rosenbaum, Larry, 65
Ross River, 73
Rottnest Island, W.A. 215
Rowell, Kathleen, Ch.23, Ch.24, Ch.30
Royal Adelaide Hospital, 41, 46
R.A.A.F., 162, 182-3
Ruined Ramparts, 93
Rum Jungle, 166 Ch.33, p.174
Ryan, Mark, 207
Ryan, Ned, 4
Ryan, Rolly, 168
Ryan, Tom, 157, 165
Ryan's Camel Party, 2, 4
Ryan's Well, 109

Sadadeen, Charlie, 15
Saint Thomas Church, 2
Salt Lakes, 141-2

Sandover Hwy, 74
Sandover Massacre, 66
Sandover River, 66, 72, 213
Schauber, Max, 81-2, 119
Schooling, 14, 32, 53
Schultz, Charlie, 117
Scott, Alex, 78
Scott's Hotel, 79, 129
Shannon, Jim, 23
Sharpe, 165
Shell Company, 153-4
Sideek, 8
Simpson Desert, 25, 29, 30, 205, 216-7, 233
Simpsons Gap, 64
Sladen Waters, 189
Slight, Albert & Mrs, 119
Smith, Bronco, 60
Smith, Fred, 8
Smith, Walter & Mabel, 66, 216-7
Solar Power, Ch.36
S.A. Edn.Dept. 13, 33
'Southern Cloud' 54-5
'Southern Cross' 54-5
Speed, Don, 146
Spencer, Sir Baldwin, 25, 29
'Spotted Tiger' Mine, 46, 101-2
Stacey, Norman, 81, 116-8
Standley Chasm, 20
Standley, Mrs Ida, 20
Staunton, Johnny, 120, 127-8
Steer, Mrs, 161
Stefansson, Valhjalmur, 29
Stirling Swamp, 103, 115
Stirling Hills, 202
Stockowners Assoc. 151
Stockwell, S.A. 2, 25, 29-30
Stolz, Pastor, 29
Stone, Bill, 98-99
Stott, Sergeant, 18, 43, 46
Strangways Ranges, 125-6, 136-8 206
Strappazon, 101
Strehlow, Carl, Pastor, 32
Stribble, Jack, 208
Stuart Arms Hotel, 37, 78
Stuart Plain, 118
Stuart, township, 1
Sullivan, Queenie, 53

Sultan, Guppa, Ch.35
Surveyors, 24, 35
Swan, Mount, 63, 65, 90, 102, 165
Swanson, Jack, 139-40
Swofield, Captain, 118
Sydney, 96-7, 130, 220

Tailem Bend, 98
Tanami, 95
Tanami Desert, 71-2
Tanami Transport, 204
Tarlton Downs, 209
Teague, Una & Violet, 33
T.E.P., 172
Tennant Creek, 2, 70-3, 78-81, 96, 103, 113, 119, 122-6, 127, 131, 133, 166-7,175,176
Thomas, Maurie, 66-7
Thring, River, 235
Tilmouth, Harry, 77
Tilmouth, Wauchope, 68, 122-3
Timms, Mr C., 76
Tindall, Maisie, 131
Ti Tree, 72, 115
Tobermory Station, 206
Toby, Alfie, 224
Todd River, 59, 218
Top End, 244
Top Springs, 169
Tourists, Ch.10
Traeger, Alfred, 37
Turner, Alf, 59, 88
Turner, Dick, 59-60, 142
Turner, Jim, 59, 66
Turner, Johnny, 233
Turner, Tom, 59
Turner, Mrs, 89
Turpin, Dick, 95
Turquoise, Ch.40

Underdown, Daisy, 95, 127, 148
Underdown, Johnny, 148
Underdown, Ly, 95, 127, 148
Undoolya Station, 31, 44
Unka Bore, 206
Unka Creek, 206
U.S. Airforce, 132
Urandangi, 208

Vacuum Oil Co. 98-9, 153

Valley Bore, 216

Wagner's Wreckers, 99
Wait, Roger, 157, 200
Waite River Stn, 63, 150
Walkington, Ben, 43, 72
Wallis Fogarty, 23, 43
Ward, Dr.Keith, 25, 29
Warner, Robby. 236, 239
Wauchope, 72, 142,
Wave Hill, 199
Webb brothers, 67
Webb, Quentin, 235
Weller, Sam, 97, 100
West's Theatre, 29
Wheelhouse, Doug, 146-8
'Whip' system, 16-17
White, 'Knocker', 228
Wiese, Ivan, 157
Williams, Harold & Zena, 82, 119
Windelman, Ivor, 77, 107-110
Windle, Mrs. 51
Windle, Phil, 23, 77
Winnecke,63, 66-7, 69-70
Winton, 131
Wolf, Harry, 23, 77, 109
Wood-gas producer,
  First, Ch.23
  Second, Ch.44
Woodford Creek, 36
Woodgreen Station, 143
Woods, Lake, 81, 117
Woods, Sonny, 186-7
Woomera, 216
Wunjawara, 88
Wright, 165
Wyndham, 199-201,
Yankalilla, 244
Young, Bert & Helen, 206
Young, 'Bogger', 155-6
Yulara, 245